Chess Openings For Beginners

The complete guide to chess openings, tactics and strategies to become a grandmaster of chess

CHESTER ADDISON

SUMMARY

INTRODUCTION

A game of chess can be approximately separated into three different stages:

1) Opening Phase

2) Middlegame Phase

3) Endgame Phase

The Middlegame phase is where the heaviest combat is generally carried out; the pieces will be whittled down, until maybe there are just a handful of them on each side, plus a scattering of Pawns, which roughly implies the transition to the Endgame stage, where either Won/Lost or Drawn will be the game.

Now, the Opening Phase, for the intent of this book.

The target for both players during the Opening process is to build their Pawns and Pieces into positions that they think will provide them with the best possible chance of winning the Middlegame war, which will then, ideally, be transformed during the Endgame phase into a Victory.

Chess has been evolving for over 1400 years, in one way or another. At that time, to build their respective forces, the various chess masters explored various sequences of

4

movements.

These sequences are collectively known as the Chess Openings (or, for short, Openings) and the good ones have been more researched and studied through the years, as the game of chess developed from a competitive hobby into a competitive sport worldwide.

With such an intensive degree of study, some of the earliest openings that once proved successful have since been dismissed by players whose survival depends on playing in chess tournaments as leading to substantial material failure, a position of serious disadvantage, or even the loss of the game - an example of this is the King's Gambit, which was announced by former World Chess Champion Bobby Fischer as an example.

All the noteworthy openings have been recorded and documented because of the importance of having a good start, in order to avoid the needless trial and error that previous chess masters once encountered. All we have to do now is review and investigate these recorded opportunities in preparation for the Middlegame war, to find sequences that fits the way we want to build our army.

The study conducted by previous chess masters eventually discovered sub-lines focused on these solid opening maneuvers in addition to the simple opening lines. "These sub-lines are referred to as "Variations". So, you will have the root opening movement(s), so they will branch out into separate sub-lines or sub-sequences at some point (one root Opening sequence can have multiple Variations).

If you will enjoy this book, I would like you let me know your opinion with a review, thank you!

Let's dive inside the book right now.

CHAPTER ONE

Chess Openings - Gaining the Advantage

One of the fun board games often played by all people will be chess. The appropriate strategy for the target will be to negotiate with the rival's king where it is unable to do opposite attacks and has nowhere else to go. This kind of board game. With many experienced players playing this kind of board game, however, there are now a range of moves possible to carry out a final blow to the enemy.

What People Has To Know

Chess is not just an easy sit-down game; it involves a careful analysis of enemy actions in order to execute the necessary counter-attacks to win. And with the various patterns used, it also has vulnerabilities that are effectively meant for it. There are also chess courses available almost everywhere. From basic movement lessons to advanced pro movements, each client is supported by skilled athletes. The users are only given little time in tournaments to perform a phase and with the stress of the game; it is difficult to consider efficiently to make the required steps.

Explaining the Moves

In this game, there are different opportunities and each has its own way of providing a special way to fight the enemy in different ways. However, if the enemy is already in the same offensive or defensive action, a maneuver that was originally intended could be wrecked and another capable move may be introduced. A professional mind is needed in this game, even for beginners who are still adapting. For this game, there are lots of tournaments and the gifted thoughts will mainly win, but often good fortune is specifically required when two professional gamers play.

The Different Types of Openings

Any step has its weak points of its own. Openings are possible to execute such attacks that will destroy the movement of an opponent. Ruy Lopez, Sicilian Defense, Pirc Defense, Queen's Gambit, and so on are some of these openings. Each step has a specialization of its own in the game. Because some are just great for a defensive role, others are great for both as well. Nevertheless, each opening varies with the sort of side the player is on, whether black or white. Due to the fact that a sequence had to be professionally made to execute it, precise moves were minimal.

Doing the chess openings correctly will cause a change in the game. However, not all gamers rely on one kind of opportunities as the challenger might probably do the counter-attack or make the defense after that. People who are also masters of this game are likely to make multiple opportunities for the prepared raid; this is done to confuse the enemy and keep the opponent from learning precisely when to move next. Some openings often sacrifice other pieces to achieve a more productive gain. This is achieved, mainly, through the use of pawns.

In this game, learning the different gestures can be an advantage. Rather than depending on chance alone, it helps to know more. Skill is what this game needs, and without it, without warning, a player will simply lose. The player who is well aware of the opening would certainly have a better chance of winning.

There have been hundreds, if not thousands, of books published on chess openings. In chess, the normal tendency for beginners is to purchase a book of chess openings and commit particular openings to memory. They normally get unstuck as the chess beginner then attempts to submit a particular opening in a real chess game when their adversary performs a non-book pass. First of all, it is best to learn good opening concepts that will then encourage you to change your openings depending on the movements of your opponents.

1. Control the Center

The first theory is to control as much of the chess board's center as possible. The main squares are d4, d5, e4 and e5, respectively. This would occur naturally in most openings for white, with the two most famous opening movements being the move of the Queens pawn to d4 or the move of the Kings pawn to e4.

2. Develop your Pieces

Second, moving the Queens or Kings pawn then controls either d5 or e5, it also offers space for the more important pieces to move. Starting with your Knights, followed by the Bishops, put in to play the second concept of forming your main pieces. In order to build your Bishops entirely, you will need to move another pawn.

It will help defend your advanced pawn by shifting your Knights to either c3 or f3 and also control more of the board center.

3. Castle Early

Then you can get to the Castle as quickly as possible, preferably on the hand of the Kings. This helps to defend the King from attack and to free the center of the Rook.

4. Connect your Rooks

You can then move the queen after you have Castled to encourage your Rooks to link and have free movement on the back rank. However, move your Queen with care, as another principle is not to overexpose your Queen in the opening.

5. Build a Solid Structure

Unless the pieces have been created, do not be tempted to strike. In the chess opener, your objective is to create a strong structure from which you can then attack your enemy. Your chess pieces would have more freedom of movement by upholding these rules and be less vulnerable to attack.

You can then learn a chess opening or two if you are familiar with these concepts and if your opponent does not obey the book moves, you can fall back on these rules as what to do next.

CHAPTER TWO

Making a Strong Chess Opening

There are a variety of ways to start chess games, but one of the more common will be the opening of the Ruy Lopez. This type of game, also known as the Spanish opener, offers White an early chance to catch one of the Knights of Black, even though it puts a Bishop at possible risk. The name of this move derives from the 16th century priest Ruy Lopez de Segura, who, among others, undertook a thorough study of this move, which he then published in Libro del Ajedrez, a book on Chess. Although the move was known until Ruy Lopez studied it, until the mid-19th century, when it was rediscovered and high importance was put on its potential as an opening, the move did not become famous.

This opening of Chess is still in general use, being one of the most sophisticated openings of Chess currently known. One of the most common openings in master level play is the double King's Pawn opening and most players have followed a version of this at some stage during their playing careers. This Ruy Lopez move has been widely established and theorized, depending on the double King opening, with several different logical openings available to both players. Any game that starts with a Ruy Lopez will offer either player a good plan to focus their game on with several different options available.

For any player trying to have a good game, several other openings are likely. At the beginning of a game, one of the most important tasks is for any single player to take control of the middle of the board. This is commonly considered to be the best position open to someone playing a chess game, and any player would have a very difficult time winning without possession of the middle of the board. Therefore, any strong opening would be constructed around attempting to take possession of the board's middle parts. This is why, in particular, the Ruy Lopez is so powerful; it positions pieces in the very center squares of the board immediately.

The chess game itself can evolve into either an open game or a closed game, based on the type of opening that is used. Open games are those in which one player occupies the middle of the board but is left open, allowing the Bishops and Rooks to pass across space with great versatility. This makes it easier to play a faster game, allowing it the opportunity to hit more frequently. The game will have the feel of a series of small skirmishes that will yield a combat winner.

However, A Closed Game is one in which the middle of the Chessboard is blocked, stationary, by pawns remaining there. This does not allow quick versatility and a slower, more painstaking style of play will always be created. The Pawns can form a stronger defensive line that is much more difficult to crack through than an Open game. In these slower moving games, players also have more difficulties and, due to this, open games are more commonly recommended.

Memorization is one of the things I cringe about in the way I teach chess. The player's chess development is not increased by beginners who are trained to memorize chess opening positions and not to grasp the ideas behind them. Let me explain to you the fundamental opening concepts of chess in this article and how you can use them to strengthen your game.

I would like to break the opening principles into two terms for the sake of simplicity. Time and Space.

Space

A player who owns more of the 64 squares on the board while playing chess has a better chance of winning the game. To allow us more freedom to cover a lot of squares, taking care of the central squares of the board is also a significant thing. For the sake of our chat, let's look at the knight. If there's a knight in the corner of the board, it looks dumb. It not only manages fewer squares, it also has fewer choices for an attack to go for the kingside or queenside! On the other side, we have a major spatial advantage to deal with if our pawns and pieces are oriented toward the center.

Time

Tempo, as it's called in the world of chess, is important in the rapid construction of an attack. If you have 3 moves in front of your opponent, for instance, then you have a lot of pieces triggered than your enemy. This is a positive thing for us because when it comes to space and number of attacking pieces, it gives us a great deal of advantage. Generally, a pawn or a piece is lost by the best attacking players only to win time or tempo. You will defeat any good opponent in chess if you take the time to learn how to do it!

Putting it Together

With this guiding concept, we discovered that without losing a lot of time, we can make movements that took care of the most squares on the board. To do this, without moving it twice in an opening, build the pieces when taking care of the central squares. You will find a change in your game if you do these easy things.

Bad Chess Opening Moves

For White to play, one of the worse first moves is 1.g4. White does not compete for the middle with this move, a significant idea in chess, and offers Black a strong aim to strike. If Black replies with 1...d5! to 1.g4. He occupies the centre and, with his bishop on c8, immediately assaults White's loose pawn on g4.

If White, for instance, with a move like 2.f3, defends his pawn, this can lead to a fast mate. Step 2...e6 appears harmless but challenges the queen with a dangerous checkmate on h4.

White can defend by playing 3.h4, but the threat of matching White on the e1-h4 diagonal is revived with 3...Bd6 Black. The 4.Rh3 move (defending against Bg3++) ends in a lovely mate in two. Black is able to sacrifice his queen with Qxh4 and Black finishes the game with 5...Bg3++ after White takes the queen with 5.Rxh4 (there is no alternative).

Admittedly, this example may be an unusual occurrence in practice, but it shows that by playing the easiest of moves in response, Black may take advantage of these poor opening moves.

Another bad chess opening move is 1.f3 and without doing

something helpful, it irrevocably weakens white's king role.

In comparison, 1.Na3 and 1.Nh3 contain bad moves. Not only do these chess moves ignore two basic rules at the same time that "a knight on the rim is dim" but also force White to move the same piece a second time in the opening, losing precious tempo, after Black plays 1...e5 or 1...d5 by capturing the knight with his bishop, threatening to double White's pawns.

DUBIOUS CHESS OPENING MOVES

For starters, 1.h4 is a questionable move. The move is not as bad as 1.g4, but it does little to help obtain control of the central squares that are significant. In comparison, after the move h4, castling kingside is less appealing as the kingside is already weakened. Thus, among serious chess players, the move is seldom seen.

With this move, White clearly can't compete for an advantage, but it shouldn't be too bad as it doesn't build any vulnerabilities.

Chess moves such as 1.e3, 1.d3 or 1.c3 can be played and do not compromise the position of White, but there is no special point in playing them except to prevent theory and the planning of the opponent. However, since we're looking at all the moves from the viewpoint of a novice, we don't need to think at all about preparation at an amateur level.

Thus, those chess moves waste time, get in the way of developing all the pieces into usable squares, and they're not competing for center power.

For example, 1.g3 is not a bad move at all and can be transposed to other openings, such as English. However, the main drawback of this move is that it allows Black to occupy the center with whatever moves they want to play.

In essence, it is advised that every beginner avoid the weird chess opening moves listed right from the beginning of their training because they can lead from the very beginning of the game to rapid defeats and disadvantageous positions.

CHAPTER THREE

Professional Opening Preparation

Opening practice, when you play against a strong opponent, has a strong effect on the final outcome.

That's why top players use 90-95% of their preparation time to prepare for opportunities. While I do not encourage you to do the same, it is still a very interesting subject. It is certainly one of the key factors in your realistic performance in modern chess.

In the lesson "Professional Opening Preparation," I would be happy to give you some advice about it:

Today, being well qualified for a tournament match is highly necessary.

Computers make the planning of openings an incredibly effective weapon.

That's why professional players typically spend 2 to 6 hours practicing for their pre-game matches. If you can't neutralize the resilience of your opponent, you'll definitely be in immense trouble.

This aspect has become so relevant that it really is an immense, independent subject that any modern chess player can master.

I will send you some useful advice about it in this chess lessons. We're starting.

It was common in the near past to play openings, which were not completely right. King's Gambit (1.e4-e5 2.f4), Center Gambit (1.e4-e5 2.d4), Bird's Opening (1.f4) and other such items were played by citizens. It was OK at the time, but now it's surely the wrong way to play.

Your next opponents will expect it from you when you use such an opening line one time and will train against it. They'll really find some bad suggestions for you if the opener is not objectively successful. You're going to start having issues, start failing, and in the future you're going to have to avoid playing the opener finally.

There is, though, a valid answer, of course. Standard openings can be played, which relate to the simple strategic concepts. Then you may not think too much about the planning of your rivals, for no one can refute the right openings.

If the World is actually spherical, no one can deny it. I hope this is where you understand what I mean. Then even Kasparov would not be able to contradict your option if you play good openings!

A new challenge is now facing us: how can you detect whether or not a given opening is good?

You should, in general, concentrate on your overall strategic awareness. It might not be that obvious to you, though. There is a better suggestion in this case: detect how many players above 2600 play this opener. So you can only believe in their knowledge of tactics and their realistic expertise.

If you see that a lot of influential players routinely play a given opening line, it definitely means it's fine. If they use it rarely or don't use it at all, so another line should be considered.

Let's consider the next realistic situation: you play a game and very easily your adversary makes his opening moves. Obviously, he is using his training for pre-game. What are you going to do then?

Oh, it's an awkward situation, of course. And commonly, a player starts

Growing tense, beginning to think hard and attempting to locate an odd pass, aiming to crack the preparedness of his adversary. Such odd moves, though, can clearly be a misunderstanding. That's why players in those conditions sometimes get into trouble.

Let's remember the bottom line: first of all, we can neutralize an opponent's machine. That's why going into a tactical variation, when the adversary is well trained, is a massive error.

Computers have strategies that are really powerful. So, often, just by observing his machine analysis, the adversary will win the game easily.

In strategic locations, an inverse condition exists. While machines are highly efficient, in positional conditions they are still not so strong. Another important thing is that there are no lines of power there. Therefore, you can't contradict your stance.

We can infer that the most important element in strategic roles is a strategic understanding of both players. This is why we do not dread the planning of an enemy.

Let's talk of another realistic situation: you've practiced for a game and are prepared to play against the opening of your enemy. Suddenly, as the game starts, an enemy plays something entirely unpredictable, something he's never played before.

Again, it is clear that he has set up this line against you in particular. What are you going to do then?

You would not feel very relaxed there if you are not ready for such a situation. You want to play the opening line, which you know well, on the one hand. In comparison, you can not grasp other lines well enough to use them.

You want to prevent the preparedness of an enemy, on the other hand. What is the solution, then?

I recommend this to you: you should not play your normal opener, instead, as soon as possible, turn to something different.

Let's more precisely discuss it. Let's assume the first move 1.d4 was made by your enemy, although he has played 1.e4 in all previous games. Typically you play the Indian defense of King against 1.d44 (just for instance). Of course, there's something your adversary has planned against King's Indian, and he's probably practicing all these lines. So, unless you want to fight against his machine, it makes no sense for you to go there.

Maybe you're going to play a line you're not well versed with. You need not be frightened of it, though, because your adversary may not know it either! Therefore, you are both going to be in the same position and the better player is going to prevail.

It's certainly better than competing against his machine set-up before the game. As I said at the beginning of this tutorial, planning for the opening is incredibly necessary nowadays. In modern chess, this is truly a different art.

A player can often choose to begin with a series of opening moves that is considered normal. Such movements are referred to as moves of books. A player may opt for an opening that is an improvisation on a standard book move at other times. The moves vary, but the objective is the same— to get off to a good start and into an offensive position from which to unleash an assault on the king of the enemy.

In chess, to put the pieces in a good position, you need to make the full use of the openings. It is never evident at the beginning which part of the board the pieces are most needed on. It is also important that efficient control over the squares in the central region of the chessboard is retained, so that individual pieces can be moved as possible, with minimal problems. Placing Pawns on d4 and e4 is the safest way to maintain control over the central areas of the board, according to classical theory.

A different technique for successful control is proposed by the new school of chess theory. Instead of occupying the middle, the idea is to dominate the center with pieces breaking down the center of the enemy from a distance. The King is, thus, put at the center of the board. This makes it possible for all players to castle in the opening or to artificially castle the King to the side of the board.

Typically, Chess openings aim to prevent the formation of Pawn weaknesses. Pawn vulnerability is a term used in the context of a Pawn island to describe the nature of isolated, doubled and backward Pawns, or clustering several Pawns together. Any players compromise in lieu of a swift attack on the location of the enemy in the endgame. Another choice is to initially sacrifice Pawns to establish a rapid attack at the endgame stage on the opponent.

In combination of chess openings, strategic strategies used for the middle game may also be used. This involve arranging Pawn breaks to create a counterplay, causing vulnerabilities in the Pawn system of the enemy, capturing crucial square possession, and having advantageous minor piece trades.

Multiple scholars on the topic of opening chess moves have assumed that it is the duty of the White set in the opening to retain and raise the advantage by going first when equalizing the game is the task of the Black set. As many players know, though, not all of the White set's chess openings are offensive, and the Black set will also be aggressive in the initial stages and take the momentum away from the White set.

Improve Opening Chess Moves - Tip 1

First comes development! The opening stage is all about mobilizing your army for the middle-game clash and maximizing their positions.

Normally, one or two opening movements in the opening are enough. Now, if you are curious what a successful sequence of creation will be, here it is (1) since they are the slowest moving pieces, knights should be created first. (2) First are the Bishops. Placed them in positions of full scope for them. (3) If no other construction move is possible, by castling, get your king to safety. (4) And last but not least, put your rooks together and get the all-powerful queen out!

This is NOT written in stone now. This can serve as a decent reference, though.

Improve Opening Chess Moves - Tip 2

Pawn grabbing ought to be avoided.

Nabbing the pawn could mean victory in the other phases of the game, especially the endgame. That isn't the case in the opening, though. Grabbing the extra pawn means you're going to fall back in progression, and being 2 or 3 tempi behind in growth could spell your doom in wide open positions in particular.

This, again, is not a rule. If you can calculate that a pawn can be comfortably taken, then go for it. But if the extra foot soldier you have is going to face immense difficulty, and if you can't calculate the end of the line, leave the pawn alone and be on the safe side.

Remember tip 1 - development comes first!

Improve Opening Chess Moves - Tip 3

Always keep an eye on the center.

For your pieces, the central squares of the board act as a springboard - offering an easier time to travel to other areas of the board where they are required. Your parts have more mobility and activity with central power.

The opening analysis is not meant to be a Massive pain. EFFICIENTLY understand how to study openings. Know which openings you want to look into. Get realistic and productive opening tips and never worry about what to do in the game at this point!

CHAPTER FOUR

Chess Openings

The first few moves to the opening of the chess set the basis for each game of chess. For hundreds of years, several of the chess openings have been called and studied. It is important to be familiar with some of the most common openings and understand the philosophy behind the moves if you want to be competitive in chess. All you need to know about the most famous chess openings is covered in this section. If the opening is aggressive or defensive, the boards below will let you know. Click to see an in-depth video until you select the opening you like to see some of the popular chess games that have been played with that opening.

Most beginners question what they should first learn. Although recognizing principles is more important than memorizing movements of a particular opening, there are a few specific openings that all chess players can begin with because they are so commonly played. For White, the King's Gambit, Queen's Gambit, Ruy Lopez and English should first be mastered by a player. A player can practice Sicilian Defense, French Defense, Scandinavian, and Slav for black first.

For any chess player, it's important to find out what sort of strategy they want to play. There's a very different playing style for each opening. When anyone chooses what sort of game they want to play, they will learn more possibilities that apply to certain kinds of games.

Adelaide Counter Gambit

The Adelaide Counter Gambit is a powerful black attack on the King's Gambit which starts with the movements:

e5 1.e4

Nc6 2.f4

3.Nf3 f5

When white play King's Gambit, they usually give up material and then start the game at the beginning. This defense turns the tables on white and gives the middle many black lines of attack. White must be cautious since there are several pits down which they can slip if they don't play correctly. The multiple options after 3...f5 are seen below.

4.exf5

This is the perfect option for white. Black's 4...e4 could move ahead. It holds black and hits the knight from white with two core pawns.

White should play Ne5 and substitute Ng5 or assault black on E4. Black must play Nf6 from here, since d5 is vulnerable to Qh5+ in black at this spot.

Moving d5 now dominates the middle of the blackboard and opens the bishop to f5 with a light square strike. Much of the white play functions here are black Bxf5.

They could opt not to move and play Qe2 for their knight. That's a weird change, and I'd hold d5. Manage the middle of the board to cause it to obstruct its parts from properly growing. The pawn is fastened so you can't grab the knight, but that's all right.

4.fxe5

That will be a major error for white people. Black will take 4...fxe4 and the knight's knight cannot move anywhere. Then five. White Ng1 Nxe5 has a substance down and a tempo down. Black has a big profit.

4.Nc3

They could be developed to Nc3 if they don't take a pawn.

4...fxe4 take. Take. This means that the two major pawns and the knight attack.

4.d4

You certainly won't often see this, but if you're fortunate you have opened one of the biggest hidden secrets in chess. The way more players enjoy chess is playing this wild variant. 4...fxe4 also remains the best follow-up and makes black in the d5 trap. White may imagine the two sides are swapping cheerleaders, but black has the upper hand as the queen takes the game h4.

My favorite opener is the King's Gambit and this is my favorite counter to the opening. Try it and probably you're going to like me.

Albin Counter-Gambit

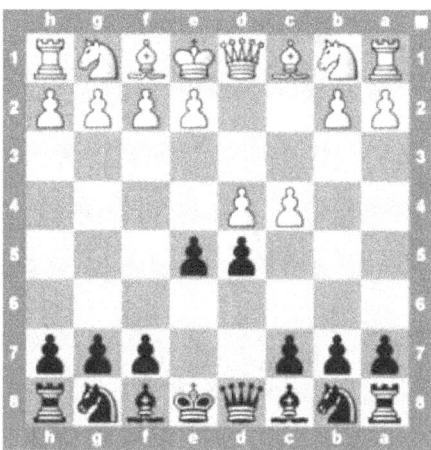

Albin Counter-Gambit is an offensive black defense against a white queen gambit, which has been so common ever since.

1. Four d5

2. E5 c4

Black gives up his cock at e5 to put his pawn on d4 nicely. This pawn is a big thorn on the side of white and several drawbacks have to be looked after from the Albin counter gambit. The Lasker Trap, which punishes white when he attempts to strike the pawn on e3 on d4, is one of the most common traps in albin counter gambit.

This opening brings a queen player out automatically and gives black several chances of combat. You obviously have to learn this opening if you are a very aggressive player, particularly black. This is also good to note for all Queen's gambit players, as your opponent will throw this at you.

Belgrade Gambit

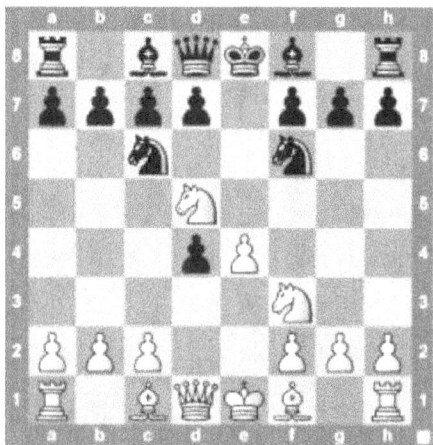

The Belgrade Gambit is an aggressive line that derives from the Scotch Four Knights Game:

1. e4 e5

2. Nf3 Nc6

3. Nc3 Nf6

4. d4 exd4

5. Nd5

White's most famous maneuver is recapturing with 5 after 4...exd4. The move to the Belgrade Gambit is allowed with the Nxd4 but with the Nd5. As several lesser known gambits, the adversary does not know how to play correctly like Black, so the Belgrade gets a lot of advantage. The good thing is that if you play 1. This variant 2. 2. Nf3 you will be able to try it because transforming into a four-knights game is very simple.

We cover any key principle in the opening, so you can play for or against Gambit in Belgrade as the most prepared individual ever.

Benoni Defense

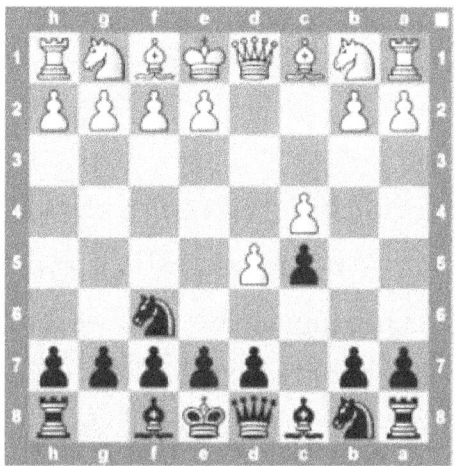

The Benoni Defense is a very offensive line which black can play against white d4's most common opening. Although many defenses against the opening of the queen pawn are closed and drawn, the Benoni Defense provides many possibilities not only to equalize position, but also to achieve an advantage and play for victory.

1. d4 Nf6

2. c4 c5

3. d5

In the Modern Benoni Defence, the main focal point of the

game is the White's center possession of the light squares and the main pawn on the d5. Black normally fianchetto his Kingside Bishop to G7 to make the dark squares some more support.

For a white player the d5 square will be pressed continuously and then used to set outposts for your minor components and to press black. You want to discourage black players from exerting control to prevent any outposts on the e6 and c6 squares. The defense of Benoni usually opens up after the opening, so that the bishops are greater than the kings, so careful to negotiate with your bishop.

Black can play a lot and have a pretty nice game when things open in the middle.

Bird's Opening

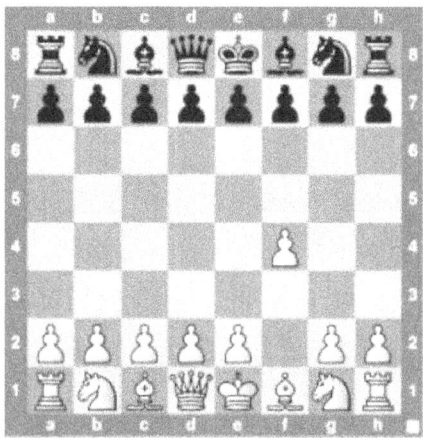

The 6th most famous opening and a rather offensive opening is the Bird's Opening. White begins to weaken his king's side and begins with his f pawn a flank attack at the middle. This opening is not used in many top level plays, but the Bird's Opening has been used to play some of the most beautiful games in chess history.

1. f4

After black defending d5 the game becomes an inverse defense for the Netherlands where white with d4 opens and black with f5 replies. White's attention is on the dark squares of the game, which are a significant deviation from the normal white light squares.

Although the light square Bishop is normally the key minor piece for white, the dark square Bishop is also nodding at this opening. White typically fianchetto his bishop to the hand of the queen and applies more pressure to the dark squares.

Blackmar-Diemer Gambit

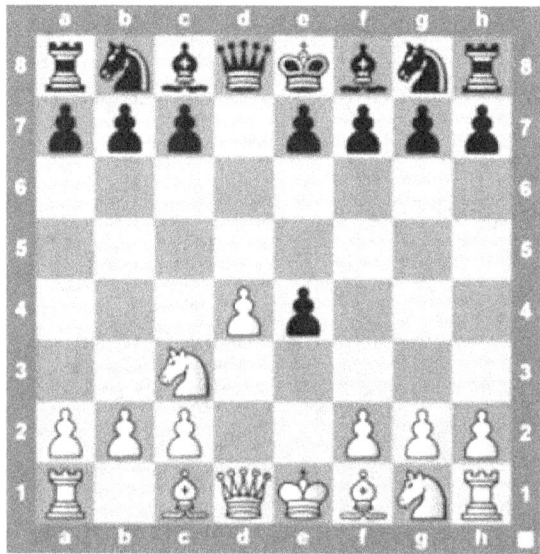

One of the most ambitious efforts to achieve developmental benefit from White is the Blackmar-Diemer Gambit.

1. d4 d5

2. e4 dxe4

3. Nc3

White looks to give his king's pawn away from the second move. While it's unsound for many top GM players, many club players have great success and I suggest playing the game from time to time if you're a very competitive player.

Then he looks at improving his knight to c3, after white offers his pawn gambit, followed by the f3. This move reveals only that players who want to play the opening like playing beyond traditional theory. After black capture on f3, white can either capture the aggressive line (Ryder Gambit) with his knight or catch a queen and allow a black queen to take a white pawn at d4.

When whites want to play the Ryder Gambit, blacks can fall into the Halosar Trap several times.

It can also be recalled that this opening can be transposed from other openings. White opening with e4 and black reacting with the Scandinavian Defense d5 are among the most common line. You can't play D4 now, and you have the same spot if you don't like playing against the Scandinavian Defense.

This opening is fun for players who don't mind giving up one or two pawns in the early part of a game for a major development lead and a chance to chase king's opponents all the time around the board. However, if you just want to play if you're up in the thing, that's not yours.

Bogo Indian Defense

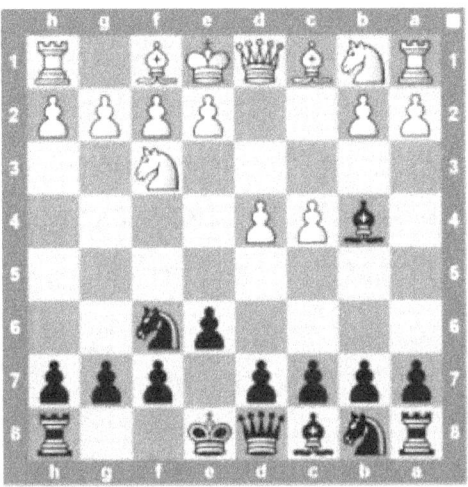

The Bogo Indian Defense is a chess opening that arrives after the starting moves:

1. d4 Nf6

2. c4 e6

3. Nf3 Bb4+

Many white players would choose the 3rd move Nf3 rather than Nc3 because they want Nimzo Defense avoided. Black has some Nf3 options, but Bb4+ is a typical move seen at all stages.

After Bb4+, White has three key options. 4. Bd2, Nc3, Nbd2.

In Nc3, the Nimzo Indian Defensive line begins to descend. Bd2 is the key line that white and I suggest should go down. It causes the black bishop to withdraw on b4 or probably lose the job. The pawn on a5 or Qe7 can be defended by Black. Both allow white to access more of the board core. The second choice is Nbd2, supporting a push to e4 and blockening the B on c1.

Bogo Indian Defense

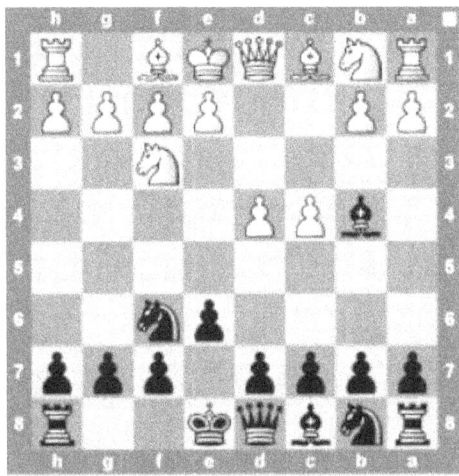

The Bogo Indian Defense is a chess opening that arrives after the starting moves:

1. d4 Nf6

2. c4 e6

3. Nf3 Bb4+

Many white players would choose the 3rd move Nf3 rather than Nc3 because they want Nimzo Defense avoided. Black has some Nf3 options, but Bb4+ is a typical move seen at all levels.

After Bb4+, White has three key choices. 4. Bd2, Nc3, Nbd2.

In Nc3, the Nimzo Indian Defensive line begins to drop. Bd2 is the key line that white and I suggest should go down. It causes the black bishop to withdraw on b4 or probably lose the job. The pawn on a5 or Qe7 can be defended by Black. Both allow white to access more of the board core. The second choice is Nbd2, supporting a push to e4 and blockening the B on c1.

Budapest Gambit

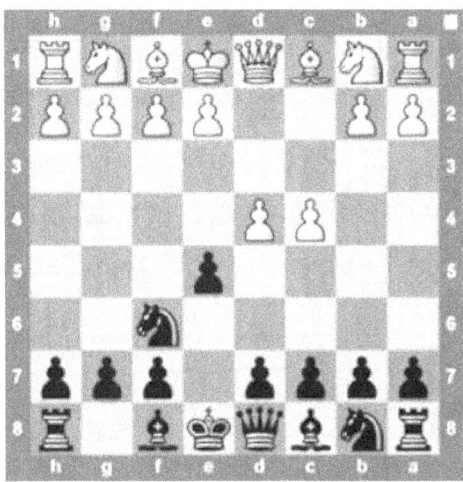

The Budapest Gambit is one of the least popular games but also has a lot of black playing. White can easily fall into a trap on the main line, for example, that ends in the game at a very early stage.

1. d4 Nf6

2. c4 e5

In the second movement, black looks at keeping his cock in e5, but then continues improving his pieces to put pressure on the e5 cock. White will never be able to retain the extra pawn and will just give the pawn back many times and keep creating pieces, not caring about the pawn gain.

In most gambits, the side that gives up the content determines the way the game goes on; it is usually white in the Gambit of Budapest to decide how the game goes on. In the main line, it is white who will determine whether he has two double pawns or the pawn and the pair is dual bishop. This is not to suggest that it is not yet playable for black, but in Budapest, white has more options than in other gambits.

Calabrese Countergambit

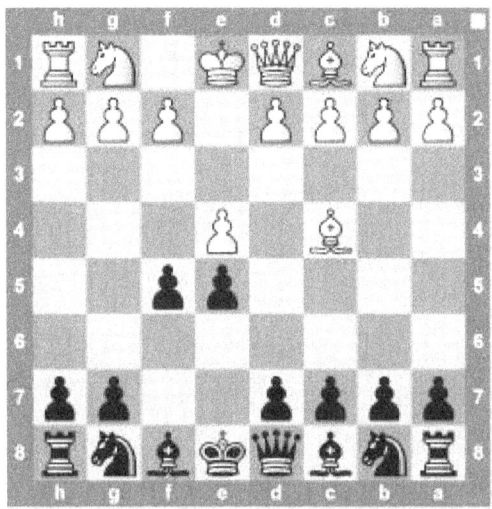

The Italian Game is the attempt of whites to establish rapidly their lightweight bishop and dominate the middle of the floor. The Calabrese Countergambit is a black way to abandon its F-pawn to foil the whole white strategy. The aggressive white wants to be at the opening at which he falls through possible traps.

1. e4 e5

2. Bc4 f5

Black generally ends up with a very strong centre, while white pieces battle to find broad squares. White has to play very accurately, otherwise he will quickly be in great trouble.

The Calabrese Countergambit is certainly something you can do if you want to strike as the black pieces.

Caro-Kann

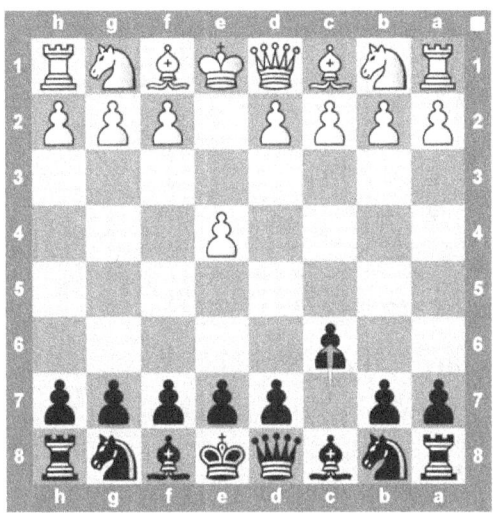

In response to the King's Pawn Opening, The Caro-Kann Defense is one of the popular openings.

1. e4 c6

Black replies with the notion to thrust the white central pawn on e4 with d5 in the next move. The Caro-Kann Defense is one of the few chess defenses in which black can gain equality on the main line and many people think that black has a better position, particularly when playing the main line at the end of the game. Typically, this is because black doesn't sacrifice the pawn arrangement and usually gets a smoother end game.

The Caro-Kann has many variants but the main line is 2. 4 d5 3. 3. DXE4 4 Nc3. Bf5 Nxe4. This is an essential framework to be seen and understood by all Caro-Kann players. A Caro-Kann opener will also transpose into a French defense but black normally has a pawn on c6 when it plays the main line. Black's squared light bishop will then come out and ultimately play e6 (usually played after the light square bishop is out so it is not blocked in). Black normally chooses to take his knight to d7 with the pawn on c6 and helps the potential knight on f6. Black will carry the queen to the c7 while his bishop in the dark square has a lot of lines and the pawns do not pause.

If the Caro-Kann doesn't suit the main lines, it's typically French defense, then if you want to play Caro-Kann I suggest that you study in French defense. Caro-Kann is neither a smooth opening nor a rather violent opening. However, the Caro-Kann is a powerful defense that can lead to a black advantage near the later stages of the chess match. I highly recommend using this in the chess games to those players who have a solid foothold on the pawn system and end game strategies.

Colorado Gambit

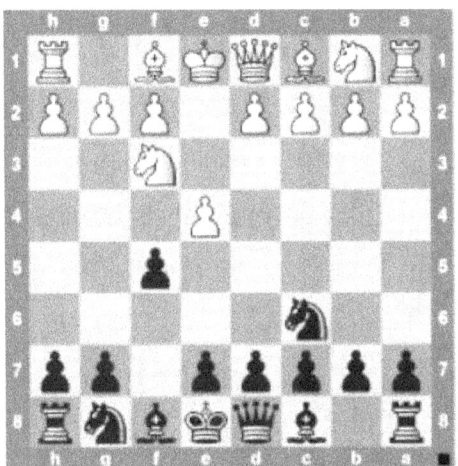

In the Nimzowitsch Defense, Colorado Gambit is an offensive line. In return for some lethal attacks on white Black looks to disrupt his king's side pawn structure. This is a solid black opening but is typically better used to surprise a player who has never learned all the lines.

1. e4 Nc6

2. Nf3 f5

White must play cautiously, as there are many traps that black can play and risky lines. In comparison to other gambits, black will rather than take pawn on f5 than white.

Instead, black would like to control with the f5 pawn the center of the board and to create the remaining pieces rapidly.

Many combinations have healthy lines and violent lines in this opening. I say you go for the more offensive lines if you want to play the Colorado Gambit. Since e4 is so popular, you can check out this opening several times.

Danish Gambit

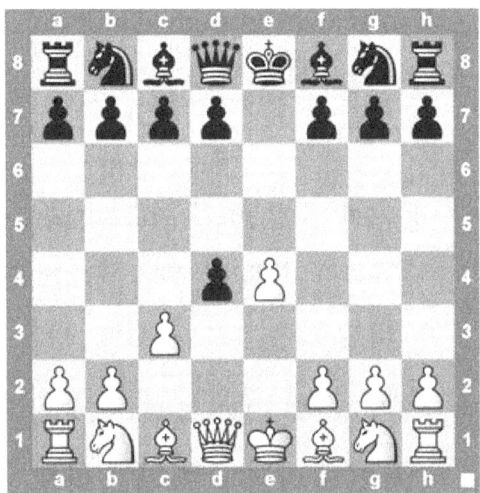

The Danish Gambit is one of the most attacking openings as white requires two pawns to be sacrificed for rapid development and attack.

1.e4 e5

2.d4 exd4

3.c3

Sometimes, you'll see the Danish Opening games in fewer than 20 moves, when white either breaks through and checks the king early or white struggles miserably and is left in smash.

Where agreed, white would have developed a solid bishop couple that stares at the black king's side because black won't develop either of his pieces. Many lines have been studied that allow black to properly defend and protect the material advantage, and black typically gives some material to recover some growth.

Whereas white loyalty to his bishop pair is necessary because the open board helps white to sustain the black pressure.

The opening is not only for the tired, but the Danish gambit could only become one of your favorite openings for people who love attacking early and frequently.

Dutch Defense

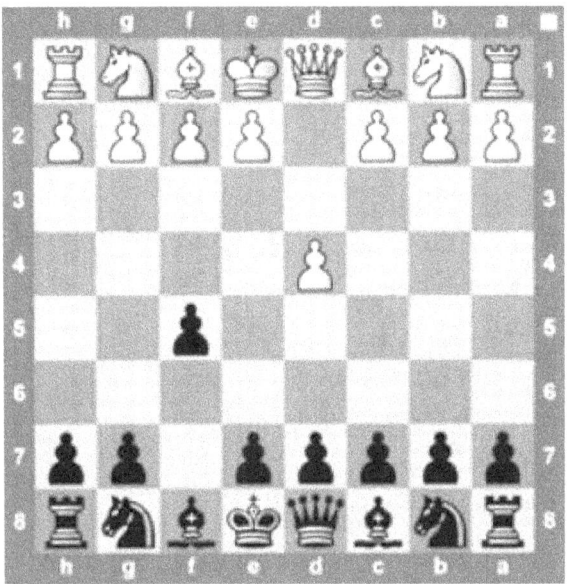

The Dutch defense in chess against the Queen pawn opening is a strongly successful defense.

1. d4 f5

Though totally unbalanced, Black looks to manipulate the E4-square. Subsequently, Black will look at his attack on the white kingside in the chess game in the future. One of the main principle is to concentrate solely on fixing this limitation, the weak F7 square that black becomes a target of white and white. In return, black has extremely active sections that are not close and can guarantee a spellbinding game play.

White normally fianchettoes the bishop of the king to G2, to give the black black e4 square support. In the king's hand, Black could even apply pressure to his bishop. As the tactics on both sides vary, most Dutch Defense games become very active and lively.

Dutch Defense offers a lot of black counterattacks for those players who always experience 1.d4 and don't want to play the Queen's Gambit line.

English Opening

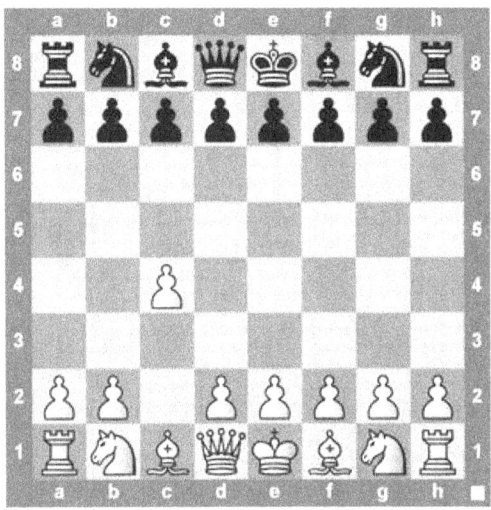

The fourth most frequent opening for white is the English opening. The opening is very fluid and always transposes into other opening lines, while the English has a very personal style. The opening attempts to place pressure on the middle d5 without the queen pawn or the king pawn committing. Due to its ultra modern style of play, this move is a flank move many players like the English (using pieces from the sides along with minor pieces to apply pressure and control the center).

The second movement of White depends on the response of black. If black doesn't try to search the center of white right continue to place still more pressure on the d5 square to give the light squares white power. Many English games start very slowly because the center is under pressure from both sides.

It's good for the English to be very flexible. With all the various openings you can transpose into, the English language can be used early in the game against any competitor. This is a perfect way to use if you prefer slow methodical games and want to be versatile.

There are also traps to protect English that are bad about the English. You should know the pitfalls you can discover, as for any opening.

Evans Gambit

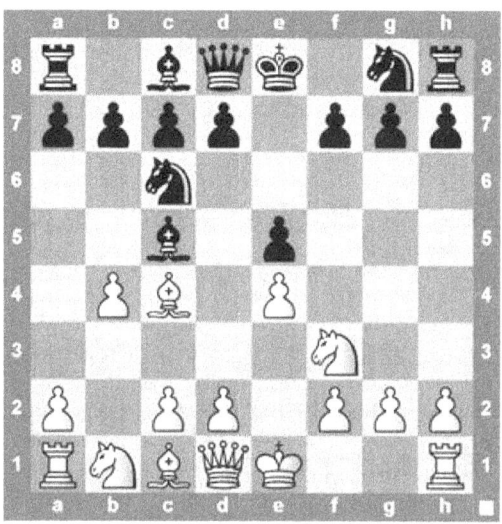

The Evans Gambit is an aggressive variation of the Giuoco Piano. For those players that like to play the Giuoco Piano I highly recommend playing the Evan's Gambit.

1. e4 e5

2. Nf3 Nc6

3. Bc4 Bc5

4. b4

Like the other gambits, Evan's Gambit leaves an early pawn to help rapidly evolve and dominate the center. White is better at Evan's Gambit than playing the gambit, and I would also consider playing it in the main lines of the Giuoco Piano.

White begins with c3, followed by d4 after black takes pawn on b4. Where a black bishop retreats to depends on the disparity. The bottom line is that Black is retreating to a 5 but you can also see that Black is relocating to c5 and that it is initially a square before he took the pawn or to e7.

You must make sure that you play vigorously while you are playing as white. This gambit is not designed for trading pieces. The Evans Gambit prevents the black king from chipping with highly active pieces in the middle and overwhelmingly black. Black can normally give back the benefit of the pawn and play counter play, but often players never want to return material once they have it. This is typically black players' weakness because white has so many lines of attack that it is impossible for black to protect them properly. A lot of games of Evans Gambit don't last long.

The major squares to be noticed by white are b3, normally the queen's position, which supports the bishop at c4. The dark squared bishop is typically a safe position for a3 and black will avoid casting of the kingdom in certain ways. The white ruler can grow to c3 as the center is controlled and no diagonals for the bishops blocked. When the king's castles are king for white, all the pieces are really busy. Black has a hard time duplicating this progress because, rather than the ideal squares, they are usually required to move their pieces in very defenceless conditions.

Falkbeer Counter-Gambit

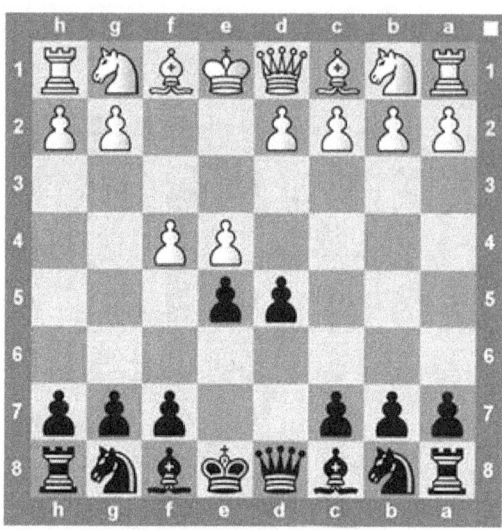

The Falkbeer Counter-Gambit is one of the quite aggressive openings for black. It is a direct counter attack against the King's Gambit of white.

1. e4 e5

2. f4 d5

It removes all the lines a king's gambling player could have plan on using, one of those wonderful things about this counter. There are still several traps white would need to watch out for because even after the second move his king is extremely exposed.

In the falkbeer counters, King's defense is one of the main items to note. Many games would not require longer than 30 shifts, since it becomes a slow one where everyone can first hit the other king. In the falkbeer counter-gambit, there are no slow developing moves. Normally, it's absolute chaos. It's a perfect defence if you enjoy chaos and you want to play really violent if you see a player that prefers the king's gambit.

Grob Attack

The Grob Attack is not a typical opening and many think of it as one of the worst opening opportunities for white people because it does not attack the middle of the board and it weakens the defense of the king. It starts by moving:

1. g4

Though this is not the best move from white, some of the desire lines exist, where white will go down and make black a hard position. Black generally responds with one of the following to the Grob Attack:

1...d5

1...e5

Where black goes doesn't matter. White must aim to bring the bishop to g2 to control the light diagonal square. White typically takes the third stage of 3 if black began with d5. 4. 3. The center control on light squares is then compromised, and if black takes the pawn on the G4, white has a direct shot on the b7.

One of the advantages of playing an opening such as the Grob Attack is that it is seldom played, and so most players have little experience with it. Although a strong player might find with enough time the best lines, all the attacking lines that white have would always be difficult. A player without a bubble or bullet chess will quickly slip into one of the many traps the opening can bring when an opposite does not match the Grob.

All boils down to planning, as in many openings. Know the Grob more than your foe and you'll definitely be way sharper than you would expect.

Halloween Gambit

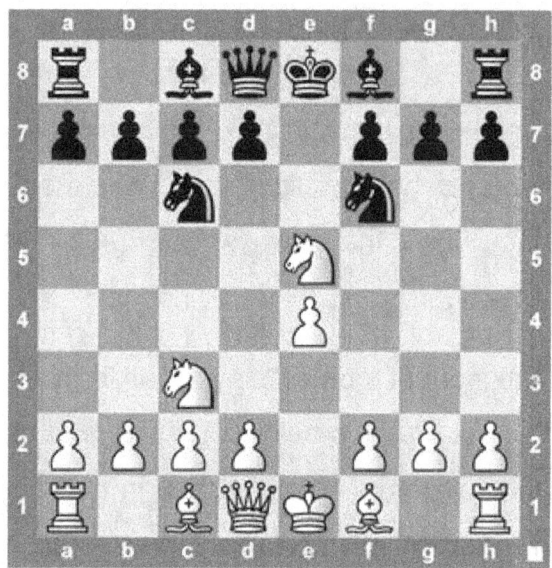

The Halloween Gambit (or Muller-Schulze Gambit) comes from the Four Knights Game, but is rather offensive, as white pawns a knight to white pawns with the intention of control the center by the white pawns. White hopes to do so as he drives his pawns ahead and pushes Black Knights behind him.

1. e4 e5

2. Nf3 Nc6

3. Nc3 Nf6

4. Nxe5

Even if black withdraw from his rulers, black has more than ample compensation for the temporary success of white in the middle. If black plays well, he should be playing an outstanding game.

However, black is often ignorant of how to defend properly, and whites will regain sacrificed material and often strike black king, who cannot deter black.

This is a worthy analysis for those players who like playing with multiple games to catch their opponents off balance. I wouldn't advise this when playing competitive, but the joys of winning with an unknown gambit often worth the chance against an ordinary chess player or even in a flash game.

Jerome Gambit

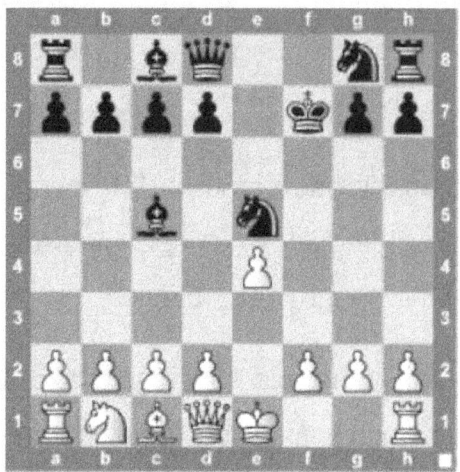

The Jerome Gambit is an offensive opening, which is typically made of the Giuoco Piano. In return for an exposed King and for the threat of an offensive attack, White sacrifices both her king side Bishop and his king. Typically movements begin with:

1. e4 e5

2. Nf3 Nc6

3. Bc4 Bc5

4. Bxf7+ Kxf7

5. Nxe5+ Nxe5

White normally holds Qh5+. This brings into the attack the most critical piece early and makes it black with the most fault. The white would have a pretty quick game if black moves Ke7 or f6. The best move for black is Ke6 but, unless you plan for this opening, it is hard to get the King to the middle of the board.

Although such gambits are totally sound and can be played as all competitive stages, the Jerome Gambit can be used for fun or lightning. If black plays right, white plays too far down and loses stuff.

This is certainly an opening for players who like super offensive openings, where you chase the King of your opponent around the board.

Reti Opening

The Reti Opening was called the Opening of the Future at times. This is because it is easy to transpose into a number of opening lines. Many players are described as being black, but black is almost expected to wait to see what the opening white line is going to take. The Queen's gambit, the English, the Ruy López or the king's Indian attack are some of its principal opening lines. As you can see, even after he has already moved, White has several different choices for choosing.

The Reti opener uses a flank strategy to take the center with a minor piece which makes for quick castling on the kingside. The most players move a peak with their initial white move to the center.

White also sets pressure on the e5 square that is black but white does not have a particular center pawn structure.

The Reti Opening is very good for those players who have a good understanding of several openings that are used to developing the king's rider to the f3 square early, as it provides more choices than other established openings.

Scandinavian Defense

One of the oldest known openings in chess is the Scandinavian defense. Black appears to weaken white power after the common e4 movement by white from the very first move.

1. e4 d5

Nine out of 10 white people are capturing the pawn on d5 and I suggest that you spend much of your time practicing Scandinavian Defense, if white is capturing the pawn on d5.

There are then two options for Black. The first choice is to recover the pawn with his queen on d5. Thus black can opt to put the queen back to a5 or to d6 after white play Nc3. It is called a poor game, because the queen came from this location. If you just want to retract your pieces if they will remain active, you should not play the Scandinavian.

The other black answer is to play Nf6 and don't automatically recover on d5. For two reasons. This is done. One black man gives up material to create a stronger core and boost the development of his small bits. The second is that white offers the material back and black gets a better plan than their queen exposed. The second is because of the white.

Semi-Slav

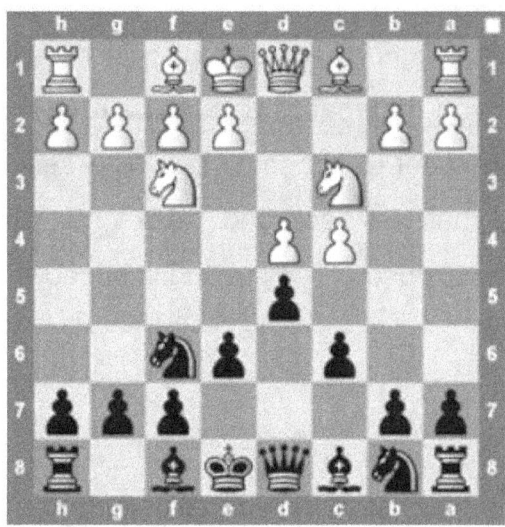

The Semi-Slav Defense is one of the most common black defenders against white Queen's Gambit. This opening is used in the Chess competition on all levels and is often considered to be one of the soundest defenses in high playing.

1.d4 d5

2.c4 c6

3.Nf3 Nf6

4.Nc3 e6

Black uses pawns and pieces to control the light squares in the center of the board, most of the time in the Semi-Slav. The Slav opening varies since the light square bishop of c8 is not formed until e6 forms a pawn structure. This gives black more time to create a firm pawn around the D5 pawn, but at the expense of its lightweight, square bishop's slow progress.

White has two key concepts on which he will usually play. The first thing is to develop his dark square bishop on c1 before closing the pawn with e3. Second, the pawn is secured by c4, and the dark square Bishop is hindered in creating e3 immediately. The complexities of the game are determined by how many times the white answers.

Black normally counteracts attacks from the queen side of the board and tries to force the light squares to dominate the middle. If black will level out, he can be much better off at the end of the game.

This is a must defend against the opening of the Queen's Gambit for those players who want the sound, basic chess.

Stafford Gambit

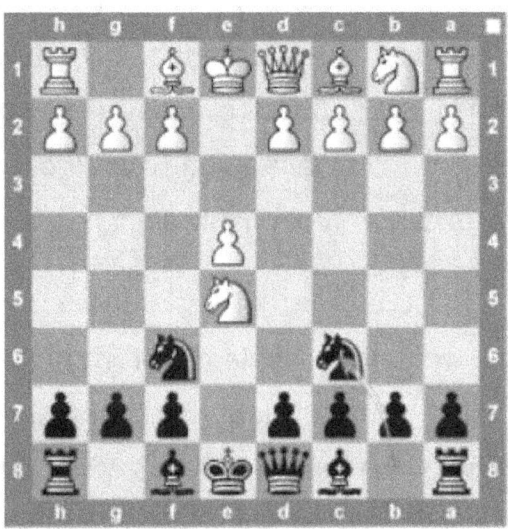

Stafford Gambit in the Petrov Defense is an offensive black line which continues with the movements:

1. e4 e5

2. Nf3 Nf6

3. Nxe5 Nc6

The accepted line continues with:

4. Nxc6 dxc6

Black offers up a pawn to openly attack the bishops and the

queen. The d4, d3 and Nc3 are generally continued by white. They all have several white traps, in which this opening is so dangerous.

5. e5

It is helpful for whiteness to move ahead and strike the knight with the e-pawn under attack. Black is pushing Ne4 now and attacking a failing f2 pawn. It's a difficult place with lots of traps. White will play Bc5 without good movements from white as they attempt to get their knight off the white with d3. Then Bxf2 + and finally the white queen will collapse if they take the knight.

If white plays d4 instead to control the middle, black may jump with Qh4. It puts a different trap. Black menaced Qxf2# but black could take Nxg3 if white defends the g3. And white will lose the rook if HXG3, and fXG3 will lose the rook and Qe4+.

If white goes on after 5 Nc3, then Black still plays Qh4, pushing the king's side to bring full pressure.

5. d3

The move aims to strengthen the white pawn chain, but it enables black to unleash a powerful 5 Bc5 attack. delf there is a development move like Nc3 is typically white, it loses to black to Ng4. There are no means of securing the square f2.

If white goes on with 6. Bg5 is attempting to pin the knight down, and blacks should skip the pin and play 6... Nxe4. When the now helpless queen is captured by White, he will lose his way. They're really in much trouble if they don't.

5. Nc3

You will see the last move 5. Nc3. Nc3. The pawn seems to be a secure move and protects on e4. The thing is black with Bc5 can also be very violent. White can't play standard center moves like 6. D4, so with 6...Bxd4, black can only take. The knight on c3, who defends it usually, is gone.

If you manage to get your light bishop on e2 or c4 black, you can start strong attacks with either Qd4 or Ng4 on your king's side.

I hope that all the traps in the Stafford Gambit can begin to be seen. While black gives up material sooner, this opening has a lot to try against your opponent.

Tennison Gambit

The Tennison Gambit is an anti-scandinavian defense offensive line which begins with moves:

1. e4 d5

2. Nf3

Black almost always captures the pawn on e4 and white has to respond with Ng5. In an interesting place, that puts black. When Black is attempting to hang onto the pawn advantage, black is going to lose his position or even the lead in the material in certain lines.

White is able to recapture and grow very well in the game if black does not hang on to the stuff. Black will come out with perfect action, so be careful to play the opening up against very strong opponents. It can be very difficult to play correctly for someone unfamiliar with this as black.

Traxler Counter Attack

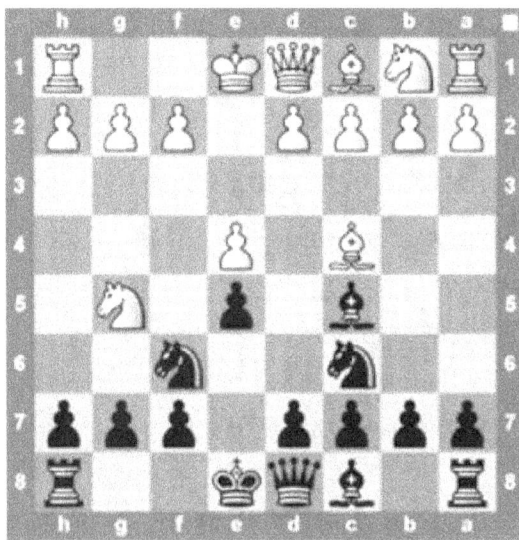

The counterattack of Traxler is extracted from two black-and-white knights' defense into the Italian game.

After his knight is switched to white, he hits his weakened f7 square and black counter with Bc5, despite his own defense for a deadly assault on his king's sides.

1. e4 e5

2. Nf3 Nc6

3. Bc4 Nf6

4. Ng5 Bc5

When whites picks up the pawn at F7, black will launch the counter-attack with BXF2 and fork the queen and the rook at the same time.

It is very hard for white to resist chasing his King around all the board and even though black is down early in the material, black can easily put his pieces into play very quickly. White generally finds it difficult to move one of his queen's side pieces.

This is an opening for very violent and imaginative players.

Are you enjoying this book? If so, I would be really happy if you could leave a review on Amazon. It helps me to understand how I can improve the book for you. Thank You very much!

CHAPTER FIVE

Basic Chess Opening Rules

For beginners, this tutorial is a must read. However, if you are a seasoned player, you can always go through the lesson and verify that you have all of these rules in mind and still obey them.

The opening stage in a chess game is very critical because it decides what sort of intermediate position it can take.

Every opening has different components – pawn structure, control of space, the creation of components, etc. At the same time, any universal theory applies to and opening, followed by every chess player and Grand Master.. There are also basic principles.

BASIC OPENING PRINCIPLES

Move the middle pawns (e4,d4,e5,d5) so you can take the initiative and have a nice room in the center.

Check or use center (the board center has the key to a

Chess game). Chess game).

Develop minor pieces (knights, bishops).

Castle your king: the king must be protected or you will begin an assault in the midlegame and you should encourage your king to be targeted. You will castle at the Kingside (o-o) or Queenside (o-o-o) – depending on your opening.

Don't move the same piece or pawn twice unless it is a major attack or you win stuff against your opponent (opening stage only).

Do not move your queen if she does not attack your adversary even if you gain material, for your adversary is able to attack the queen quickly with his little pieces, and you will have to lose some time in taking back your queen (opening stage only).

EXAMPLES:

Example – 1

Let us look at a game that is a good example of how to build little pieces not move your queen– as I stated earlier in the basic principles of opening

In the above diagram that follows movements 1. e4 e5 2. Qg4 d6 3. Qh5 Nf6 4. Qf3 Bg4 5. Qa3 d5 6. Qa5 Nc6 7. Qa4, White just had his queen established, leaving him behind Black far behind. In the meantime, Black has well followed the values of growth, playing three small parts and bullying the queen of White endlessly.

So, though White moved and proceeded to move his queen

early in the opening point, Black formed three small pieces. Black is much easier in the opening here and is able to take the game comfortably.

Example – 2

See a particular example of how the center is being filled or controlled.

In the above diagram, White was outstanding at setting up the center power. His cockroaches at e4 and d4 control several main squares, while his cockroaches at f3 and c3 can easily be saved to the central places and even eliminated if it is needed.

In comparison, the first few movements have been played terribly by Black. His pawns on a5 and h5 have no leverage or control over the middle and his cavaliers on a6 and h6 are not well positioned.

Example – 3

Let us see now as I mentioned above in the fundamental principles of opening an example of king castling.

Both players cast into the first five actions of the game. In the map above. Both kings are healthy, and neither player must be afraid of a swift fellow here.

It should also be remembered that the stands around the kings — in fact, the three pawns before the castling kings — were not interrupted.

If these peones (g4-h4) are moved in the opening, the king is usually very weak as he opens attack lines for the pieces of the other player. The safer you are, then, as soon as you castle.

Example – 4

We now evaluate a game with some simple concepts for opening. This is where the game is going: 1.e4 e5 2.Nf3 Nc6 3.Bc4 Bc5 – all the players are now moving central pawns to small pieces.

4.c3 Nf6 5.d4 – White dominates the center on d4 and e4 squares with its central cobbles. Bb6 7.d6 – The center of Black has now been lost and the main cobbles of White will develop two other minor pieces (Knight and Bishop on b1 and C1) and his King's castle throughout the subsequent movements. exd4 6.cxd4 Bb6 7.d5

Black's ruler on c6 is running and it would lose a move as

the bishop has already done in his return to b6. Black still has to establish his bishop on c8, but has lost his middle.

Picking the Right Type of Opening

Considering Naming Conventions

On a little secret, I'll let you in: chess players suck at naming meetings. Chess players should be rational and able to reason and think consistently, but you would never know it if you measured them by the name of their openings.

A provided opening can be named after a person, place or thing (the Stonewall Attack is currently an opening). Few openings are named after animals, but none is named after minerals or vegetables, as far as I know.

Name Conventions can vary from country to country, but you can find the explanation and the rhyme behind them even though you adhere to the opening names used in the United States. Sometimes the opener is named after the first player who played it; sometimes the player who popularized it is named after it; and sometimes it is named after the city, state, or country where it comes into play.

Taking the so-called indian opportunities into account. English players in Calcutta came into touch sometime around the middle of the 18th century with an Indian player by the name of Bannerjee Moheschunder. At first he did not know European rules, but in a small period of time he appeared to be a reasonably effective player. He was used to playing an old game where only one square was moved by the pawns.

It would defend against 1.d4 with 1....Nf6, either 2....g6 (known as the Indian defense of the King) or 2....b6 (known as the Indian defense of the King). It would defend against 1.d4 and 2.. The Indian Nimzo (2....e6) and the Old Indian (2....d6) defenses are now in place as well.

These improvements were unacceptable in the later movement of the 19th century and early 20th century, the classical philosophy of opening play. In the 1920s, when these ancient moves began to play a talented group of players, they were called the Hypermodern School.

Often the name of an opening is based on a White move. For starters, 1.c4 is known as the English Opener. A Black move also gives the name of an opening. For starters, the French Defense is called 1.e4 e6. All openings will connect in various ways, regardless of the name of the opening. These divisions are called variants and sometimes often have names.

Variations are sub-sets of the corresponding opening. The moves 1.e4 e6 2.d4 d5 3.e5, for instance, show the French Defensive Advance Variation. The variant is called Advance, but it is also called the French opening (see figure below).

The same name can be used in more than one opening in order to further confuse matters, but the positions are distinct for multiple openings.

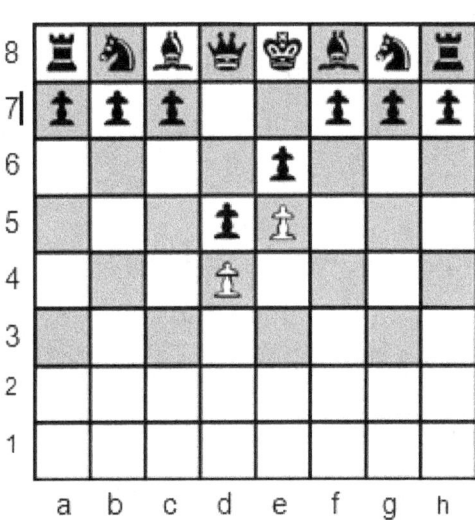

Examining Opening Types

fFor group openings with certain similarities, a category oopenings are used. These classification are dependent on the pawn structure and versatility of the components. However, chess is a complex game, and what begins as a guy immediately becomes a different one.

These kinds of games can also be sorted into general groups where those features are more frequently than not common. It allows you to create a viewpoint from which all the parallels and distinctions of the innumerable chess openings can be understood. The sections below explain the various opening categories and demonstrate what sort of player in each category is eligible.

Breaking open the board with open games

This word is used for openings beginning from 1.e4 e5.

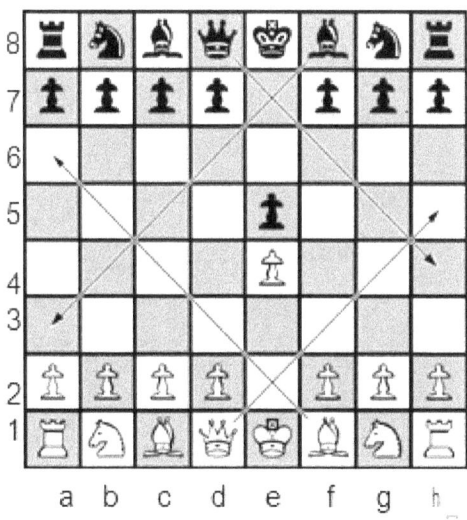

For their queen and one of their bishops on the very first stage, both sides open (the word "line" is extended with ranks, files or diagonals). At the start of the game, these lines are closed. It increases the mobility of a piece, and the strength of a piece is related directly to its mobility.

Open games can lead to fast pieces and quick attacks and fun tactical combinations. Oder they will lead to a sequence of piece trades that make the game simpler and speed up the start of the game.

There are no promises in chess, and one aspect of the game is that the rival won't play the same style of game you play. I have played so many games to count where one hand is urgently attempting to open lines and the other is trying to keep them closed as desperately.

In general, you want to open lines if you want to strike fast.

Semi-open games counterattack

For games in which White begins with 1.e4, and Black responds with everything but 1....e5. These games are by definition defenses or counterattacks. The semi-open game style disturbs the symmetry present or retained at the beginning of the game by replying 1.e4 with 1....e5.

Many of the half-open games are planned to strike White's pawn. The defenses of the French and Carol-Kann brace for a move ...d4, both with 1....e6 and 1....c6, while the Scandinavian defense forces (1....d5) and the Alexis defense forces (1....nf6) attack on Black's pawn.

The Sicilian Defense is the most common semi-open game when Black plays 1....c5 (see Figure below). The goal of Black is to construct an unbalanced game. Symmetrical games can make forming plans reasonably straightforward.

It will be tougher for asymmetric games to do so.

In any case, you are told that, regardless of whether you wish to strike White's pawn on e4 or attempt to build unbalanced positions. You see very few people who take semi-open defenses from "you go your way and I'll go mine"

Semi-open games draw players who don't care about giving away anything to make it fun. The bishop, for instance, always has minimal versatility in French Defense on c8. It's pretty much the ornery opening, though. Its first stage, 1....e6, is White's contest of d5 power. Your second move, 2....d5, White's pawn contests on e4!

In order to be in chess, you usually have to give something away. If that is your position, then what you want is one of the openings in the section on Semi-Open Games.

Shutting down your opponent with closed games

The word 'closed' means games starting at 1.d4 d5 (see Figure below). White is having trouble in playing e2-e4 because of the Black opening move. If White can handle it somewhere, the game could be opened. If not, the pieces would be able to maneuver on fewer sides.

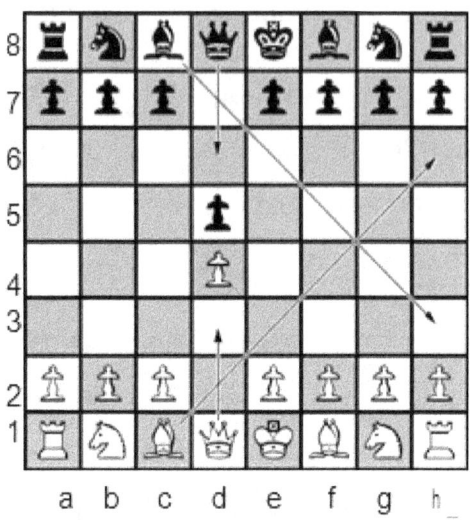

The emphasis in closed games is on the efficient development not so much on the rapid development of the pieces. For smaller component squares and fewer ways to hit the right positions, the prize is paid to get the pieces as quickly as possible. It can take some time to build a sense of how to do it. This some times involves a very subtle maneuver.

Many chess teachers therefore advise new players to start open games. The aims seem to be simpler and the movements of the pieces are more comprising. Hey f7 is the weakest square in Black (only Black's King guarded it at the beginning), so let's attack it on c4 with a bishop!

Closed games are awarded for persistence, but if you are the sort of player who will delay immediate compensation to fulfill your ultimate goal, you could have closed games.

Openings like the weakening Queen's Gambit and Slav defense are high-profile and strategically nuanced and call on the player form of long-term strategist.

Playing coy with semi-closed games

The semi-closed phrase applies to all 1.d4 answers except 1....d5. This include the Indian defenses I listed in the book before. For Black, 1....Nf6 is the typical first move.

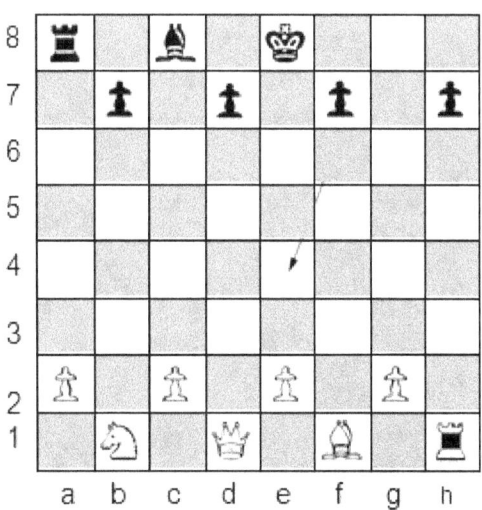

Black discourages White, but may later permit, from playing an early e2-e4. What is popular in the semi-closed games is Black's ability to allow White into the middle with cobbles in order to smash them later.

At the centre of what became the Hypermodern School of Philosophy in the 1920s was the notion of delaying pawning and attempting to manipulate the middle from a distance.

Half-close games can be hard to master, often posing the risk that White's moving pawns will easily overpower you. But these openings could be the right match for you if you like to counterpunch. As far as pawn centres is concerned, the bigger your position is, the harder it goes!

Playing on the sidelines with flank games

The word flank refers to all openings where either e-pawn or d-pawn do not advance on the first step. The 1.c4 English Opening is the most common opening on the flank by far.

White typically has one of the bishops to fianchetto at least. The term fianchetto means fianco, which means flank. White is attacking either the c- or f-pawn at the middle and attacks with pieces from a distance. Black can set up a solid pawn base, but White is determined to destroy it.

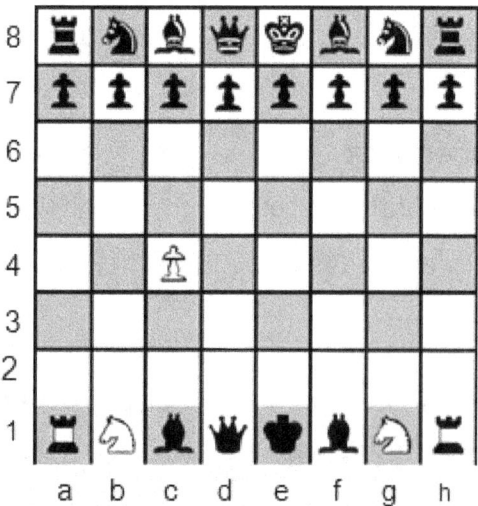

The occupancy of the middle in flank openings has been postponed, but it is not abandoned. White only likes to grow and cast a few parts safely before taking part in center hostilities.

The flank games cater to the sort of player who values what is secure and familiar because White is usually evolving in the same manner, irrespective of what Black does.

CHAPTER SIX

Taking Charge

You have control over the center is extremely vital. For several purposes, control of the center is needed; king's security and piece growth are the most important ones. It is also clear that this principle is related to two other ideas behind chess strategies and thus that when you play a successful strategy you effectively base the other strategies. Before you take ownership of the center, you must understand the present state of the center. This gives you enough time to think for the next move. The center should preferably come into one of these categories:

1. Blocked Center:

A blocked center means that it is impossible to open the middle. The idea behind the development of a blocked center is to remove any chance that opponents will triumph over it and to deny them central locations. In the 1800s it was believed that it was an indestructible idea and the best way to fight over a center order in your opponent.

Since then, a modernist viewpoint has gathered traction on the assumption of weakness. We can see this as real in Black's position in the Grünfeld Defense, which allows the

White Player to look out over the middle first, and then see Black effectively bypass it.

(Griinfeld Defense: Blocked center bypassing Black)

The below are few tips for repairing blocked centres:

• Learn how to better plan your jobs. You gain space on your territories by means of pawns to dominate the centre.

• If you choose to generate one, attempt to manage vital squares via these advanced pawns so that others can take advantage of the fun house.

• Focus your pawns and pieces to chip away if your competitor has left you with one.

• After organizing the pieces, start building them along the board flanks.

• Be prepared to put your parts in any file that might be accessible during the game.

2. Open Center:

An open center is located opposite the blocked center. It offers players an equal opportunity to grow around the centre. It is thus necessary to use the open center situation as far as possible. There are a couple of tips for dealing with the open centre: Place all the active components in the centre. In addition to helping you recover it, positioning your active elements in the center will help to unleash an offensive attack on your opponent. Do not waste time on frivolous stuff, including an isolated pawn, when building a solid midfield. Often it is good to trade a middle pawn for power.

3. Dynamic Center

A complex center is a situation in which all parties don't know who the center is keeping. This difficulty triggers unparalleled and unexpected turns in the game. To avoid this misunderstanding, the following markers must be taken:

• Clarify it first. Know who has more an advantage in the centre. You should plan for the next steps in this understanding. For example, if you know that your opponent has the upper hand, you should be careful and play more defensively. In the other side, if you realize you are holding the centre, look for an aggressive assault.

Taking a look at points numbers, component versatility, material advantage and superior pawn structure to help determine who has more power. You must consider the more nuanced advantages each player has, or it becomes overwhelming. For example, if you have a superior pawn structure and your opponent has a greater flexibility in parts (take certain moving parts into account in their background) and material advantages, even with one single piece, it's your opposite individual who has control. Their wide-ranging flexibility allows them ready to strike anywhere, but it can take you many times to place the pieces strategically.

Never carry out an attack without explanation. If the center remains your opponent, there would be a good chance of costing you an important piece if an offensive maneuver occurs. Remember then to do so.

Take out the stuff to counter, while the adversary is securely anchored to the center of the board and continues to advance.

King safety

Any chess player claims his king's defense and win the game. The concession on the safety of the king assures your game loss. Thus, the safety of the king is beyond question our prerogative at any time of the day. A few hints to make sure your king is safe are listed below: To your reference:

• Make sure there is still a wall of pawns in front of your king. Certainly the king will be easiness due to the lack of pawns or a weak pawn shield.

Arrange enough sections next to your king to defend the enemy's offensive assaults.

• Make sure you keep the middle. The center needs to be

firmly gripped not merely to guarantee the safety of your king, but also to initiate an aggressive attack on your opponent's King.

• Learn how to castle your king at the outset of the game itself. Castling is an easy way to defend your king. Again, you can remember the various openings, that help you strategically move your pieces while allowing your king to use a rock to create a castle.

If the king's capture was not a failure of the game, the dynamic of this piece's military movements would grasp in a whole new way. At first, it can sound meaningless and even dumb, but encourage me to do it. Since the adaptation of the game we know it today, the king has always been able to move one square in any direction. It essentially establishes a forcefeld around itself to shield itself from preventing opposition such that the gallant, however rude bishops, queens and rooks cannot disturb their personal space without offering protection to protect themselves.

During late game classes, however, a concentrated role, the king's hidden force takes advantage of this in order to restrict the opponent's approach and alter what might seem a weak tendency.

It's called to take the king for a walk and the king must be used to create a good and sophisticated strategic plan for the game, like every other piece of the chessboard. That is, when the king's norm shines really, get him ready to give up his walking shoes and take a stroll.

Duality of the Queen

Amateur players, with all their tempting tasks, easily fall in love with the power of the queen. She is by far the most prevalent part of the board, so the importance is nine. Until its path is blocked, it can pass the queen in folders, ranks, and diagonally at any length she wants. In truth the only trick for which she wasn't blessed is that the knight would catch a funky swoop backward. However, it is worthwhile, that the queen only took one move to hit each square diagonally adjacent before the pawns before 1475.

Since her reception, she became a demolition expert, rendering the board an intimidating distraction. For this cause, amateurs are so hooked to it that the game costs them. In their positioning and defense, they are so excited that they often ignore their other pieces and invest in the strength of their tactic, on which a more skilled player will certainly capitalize. A beginner will immediately lose faith and struggle for it in their gameplay, as he has wasted his most valuable work. It will show that experienced players do not mind removing the piece by studying other chess books and players to play a more docile game.

Therefore, the Queen's governing force is also her weakness. Too early in the game, she would hurry to seal her future and to keep down the true strength of her scheme. First of all, attempt, with a reckless and unplanned conduct and to make her one of the last parts you do, not place your queen in the precarious role.

The easiest way to take advantage of the prominence of the queen is to put your minor parts first and keep the queen on an advantageous spot deep within your own territory. If you intend to finally bring it into a position she loves, make sure the other little parts cover it well. You want your Queen as the strong arm of a leading attack to see the rest of the pieces ready to play a part in the action. So the entire degree of ferocity of the Queen can be maximized.

The Enforcer - Rook

This is a suitable title for the second most powerful work and is all too much ignored due to the influence of the queen and his launch. Many occasions, until the endgame, or even before the rook is over, after a noisy pawn a rook would not be taken out or happily played. The value this piece plays is to be understood and that the long arm of the rook supports the rest of the crew by commanding an open register as a good supporting weapon.

The Rook was the key piece of the game until the Queen

became king in the 15th century. We can see this second move behind the Queen in its point value of 5. This is currently an optimal space position for the rook on the board. Get your rooks to open files whenever possible, particularly in circumstances where the back of the opponent's base is easy to access.

This means that you have to lead with your small pieces after opening to clear the moving room for the rocks behind the lines. View locations where one of your peers can quickly give up or take an enemy piece to open the file to get your rook ready to wait. If you move the castle you will find the rook in the middle, and cover your king as a nice place for action. If you do this, move the queen to a strategic location such that rooks are linked and strengthened.

You will double or even triple this file if you placed a rook in an open file by positioning the queen or other smaller sections on it so that you take complete advantage of your adversary with an attacking ram power. Via the enemy lines to the rook in an open file during the later midgame and into the finishing line, the adversary has some nightmare. The scale of the river rocks, particularly in next rows or files, functions as an electrified space in which all the presumptuous parts left in the way are corralled by the movements of the opposing monarch. As for any cognitive tactic, maintain the permanent power of your rook by keeping it secure with another piece as far as possible.

The Chevalier

Respect or dislike them. While it's completely difficult to not know what to do with this, the tricky sticker usually derives a pronounced feeling from experienced plays to amateurs. Of all chess pieces, the rulers have the most distinct tendency to leap, both in pattern and in the capacity to jump over other pieces, because of their centuries-old appeal and the difficulty of making full use of their strategic strength.

In the overall strategy, you first need to decide the bits you will attack and which you will leave for good security. Soon you can know the squares are important to both sides by practicing, within the opening movements and in the midgame. The piece consistency section tells you that poor or critical squares are important for the whole context of the game, and in fact so important that in subsequent games you will spend much of the time choosing which squares are added at an early point. Then play them well in the midgame. Besides, you will have to learn how to either grab certain areas to advance or prophylaxize, blocking them in order to keep them from being taken over by the enemy.

Now that you have matched your Midgame strategy and determined the squares are critically necessary, you can determine whether to move your knight into the dens or retain him in order to protect your king. You must brace yourself to see every possible position to which a knight can pass, whether it is occupied or not. Learn it on both sides of the table for every pass you play and watch. This fundamentally helps you to consider the nature of the knight and discourages unwelcome plays from shocking you. Many players at the beginner and intermediate stages do not take this into consideration, which is why they fear this guy and have such reduced confidence.

When the knights can move, they can see where they do not move (game boundaries, areas packed with their pieces), where they typically do not go except when protected by a defensive advantage (blocked squares or opponent squares) and when they can go (squares occupied by an unprotected enemy piece and all other open squares). Knights in the consolidated position at least the third position have a total potential of 8 squares to defend. Any of these perfect squares are, however, blocked by one of the above instances much of the time.

A knight positioned in the first and second ranks is constrained in the possible quadrations; it does not give the potential for pull but allows the placement poor and effective. So if the knight is to reserve a defenceless tactic in order to avoid the adversary from progressing into any significant positions guarded by the knight, note this. This strategy may be advantageous, but often limits its potential.

Color-blind Bishops

Although almost any piece of chess can only be found in light and dark areas, bishops have been confined to their own color since the start of the game. Every player has a light-squared bishop and a bishop that makes a major gameplay difference at certain points. This does not identify a shortcoming for the bishop's work. In the contrary, they can have excellent long term bolts, attacks and defense for other mobile parts in a diagonal position, which is difficult to see between pillars of pawns and other parts when still covered in their own territory. This gives them the stealth of the queen and rooks that make them appear higher. However, the combination of their features just positions the importance of their piece at 3. The bottom line is to understand that in order to excel in the role of bishops you must be able to assess them in three different groups of categories, positive, bad or active.

Good Bishops

Provided the attributes of the bishops, when one is free from central cobbles, one determines to achieve. This ensures that all your core pawns have color squares opposite that of your bishop's direction. When a pawn or an enemy has a pawn chain with a bishop's color, its powers are neutralized for a long period of time.

Bad Bishops

If the preceding condition is that the key pawns have their color to block whatever effect they might have, a bishop is bad. For a bishop between midgame and endgame, this situation is particularly difficult since only one of its kind is left at your hand. Set-off positions in locked pawns, in particular in groups, can be as challenging as long distances from heavy traffic. If there are few bits left and there is a bishop or the opposite, the king will soon be moved to the opposite color square.

Active Bishop

If an active bishop is in a position to have many multiple agile possibilities, what makes him as active is what an active bishop is; the other way round is true; an inactive bishop is one with very low mobility options and 0 near how he begins the game. While a weak bishop basically has no choice other than their restricted movement, multiple options exist for an active bishop who can reverse the momentum and give his side a gain. By definition, good may be active bishops (all their main pawns are based on opposite squares), but their movement can also be constrained by the pieces of their opposing actors, and nothing can they do about them. Active bad bishops will have pieces on the same colorful squares as a same bishop, but the active bad bishop presumably will be on the opposite side of the attack on the enemy's lands instead of stuck behind the same lines. Furthermore, some projections

Pawn Structure

In 1749 André phillidor believed that the pawns were the 'soul of chess.' Until about 1909, when the legendary chess player Emanuel Lasker explained this, any chess master now takes it into consideration with considerable seriousness. This comment was not appreciated at all.

The careful placement of pawns should be understood, which will help you determine the rest of your moves and the whole strategy. These are the first pieces of the game (except a piece that often is willy-nilly) to be moved, so be especially careful as you put each of your cocks in the game, for they may only be undervalued at the cost of yourself.

Their blocking and restricting forces are a few of the pawn's attributes. Although they move very slowly one step at a time, they are effectively the hidden game's bosses. Pawns build roadblocks, halting in front of the enemy's pawn so that its efforts and structures have to be pushed by the opponent into another section.

They often avoid an enemy's more moving parts from

requesting important squares in the center of the board. Furthermore, the deliberate expansion of a pawn line controls the number of places a player can demand on the board, an important part of the mid-game strategy.

You have a strong basis for the pawn. That is significant. A strong pawn scheme will aid a lot in the execution of the plans. Any guidance on how to strengthen the pawn's structure is:

Do not separate the cowboys. Isolated pawns are easy saving goals and not an initiative.

Make sure you have a complicated pawn structure. If your opponent tries to run for his money, it is important to get a complicated pawn structure.

Note that your pawns can balance and support the movement of the other pieces in their positions. Don't let your pawn block an active piece, because your strategy can be very expensive. This refers particularly to the paths of your bishops.

A basic rule is to use your cock to patch the cockroaches of your opponent wherever possible. This forces the lanes of the versatile parts into grouped roads, which you can use by fitting out the correct parts.

Stop the islands of the pawns. When you've got a lonely

group of pawns in the middle of the board, you create holes in your defense. The pawn chains, on the other hand, are weak at their basis, the only position not secured by a pawn (a diagonal line of bonded pawns). The more pawn islands you have, the greater your enemy's attack points. Therefore, while you would like to delete pawn islands that invite gaps in your defence, you do want to note that your opponent's pawn chain is still attacked on its basis.

Advancing pawns are appealing and disadvantageous. If you have a pawn that prevents the mobility of any other pieces, it might be a clever idea to drive a pawn forward, even if it means losing the progression of the most striking parts. For advanced pawns, files and diagonals also have to be opened to take over previously inactive pieces. At the end of the game, whether you have a pawn in your opponent's sweets, you want to encourage it by all means, so that it can be a queen or smaller piece, and the dynamics of the game radically change.

One tip is that pawns frequently block an enemy's pieces from moving in midgame positions. Pay particular attention to the unpopulated areas and some of them hold a considerable strategic power for the player who can gain from them. Making sure you don't give up any of these wonderful foundations as you move a pawn along.

FINDING CHECKMATE

Pawn Race: Since both sides only have kings and pawns, it is interesting to count how many moves each side can take to promote them. Anyone who gets there first will typically win the game!

Checkmate in Two Moves: Checking at all the checks on either side is useful and seeing whether on the next move there are checkmates or other strategies available.

Don't Just Check: It's great to strike the king - but it's not always the best pass. Patiently be. Make sure a search serves an aim, such as helping to mate or avoiding the castling of the enemy. It's actually not worth it if your enemy has a valuable maneuver to get out of reach.

King Safety: Never forget your most valuable piece, the king, to defend. If there's a checkmate, the content superiority doesn't matter. Be careful to cover prospective checkmating squares near your king and, if possible, offer your king escape routes.

Weak Targets, f2 and f7: In the starting position, the f2 and f7 pawns are only defended by the king, so they can be subject to attack. If you can attack the f-pawn of your opponent with two pieces while it is only defended by a king, you can always win equipment or checkmate. The castle is a good defensive idea so that your rook defends the f-pawn.

Back Rank Checkmates: Castling is a smart idea, but it could be prone to a back rank checkmate from a rook or queen if the king is on the back rank without other pieces to protect it. It's always a smart idea to move a pawn ahead of the king to create an escape square to stop a mate.

The object of the game is to checkmate the King's opponent. When the King is put in check and does not get out of check, this arises. One of the easiest ways to develop the chess game is by learning familiar patterns that occur in games time and time again. By constructing this pattern recognition, by identifying basic patterns in more complex positions, you can begin to see more possibilities in your games.

What is Checkmate?

Checkmate takes place in Chess when you or the King of your opponent is in check, the King is unable to move, and nothing will catch the check delivery piece. Checkmate also means, considering how many pieces are left on the board, that the game must come to an abrupt end.

This is a pawn delivering a checkmate, even though the board is nearly full of bits of chess. Since the Queen defends it, the King is under immediate attack, does not pass, and can not recapture the pawn.

Two Major Pieces Back Rank Mate

This is by far the easiest potential partner in the Chess game, which generally takes place in the late stages of the game (i. e. endgame). The stronger side slices King's opponents with one big piece from the 7th rank and offers a friend from the other. This mate is very popular to know and to be conscious of, and hence essential.

Note: If the King of the opponent is trapped in the center of the row, the stronger side can "walk" him to the 8th rank by giving interchangeable checks with the rooks to the seen position, and then mating on the back rank.

Two Pawn Checkmate

This is a very simple endgame, where there are two pawns on one side and none on the other side. Black has nowhere to move in the following positions: d7 and f7 are guarded by the King of the White; d8 and f8 are guarded by the e7 pawn, which can clearly not be captured.

Back Rank Checkmate

This is a typical example of the back rank checkmate, which is a very effective weapon used as a serious weapon or a hazard by both players. His own pawns on the 8th rank (i. e. back rank) are blocked in the position above Black's King and any check by a major piece on the back rank will turn out to be deadly. The back-rank checkmate may not look anything like our first case, but it is very close in pattern. Although we're still going to use a big piece to bring a checkmate to the edge of the board, the King's own pawns are stopping him from avoiding our attack this time around.

Note: Even though no immediate checkmate is open, condition will shift very easily and it is always a safe idea to establish an escape window in the endgame where major pieces are present, you should still be conscious of the back rank threats.

Tip: In the Middle Game, there are a number of back rank mate instances as well.

Diagonal Checkmate

It was proven that the Queen + Bishop configuration lined up on the same diagonal was very effective. In this case, because black dominates the so-called long (a1-h8) diagonal, the setup is even more strong. Mates in white with 1.Qxg7#. The Diagonal Checkmate should be in any realistic chess player's arsenal. Around the same time, whilst on the defensive side, players should be careful of this kind of setup.

Note: If White's Queen and Bishop are lined up on the b1-h7 diagonal, it is conceivable to have a common form of mate, double attacking the vulnerable h7square that the King alone always guards.

Smothered Checkmate

When an opponent's king is unable to move because it is blocked by its own pieces or pawns, it is a mate provided by the knight alone. Often, this mate is very difficult to see since players don't usually expect the knight to be a mating piece. Note: When you see that the king of your opponent will not move because of surrounding pieces, you can press in your mind on the smothered mate theme.

In situations where a king is too well defended for his own benefit, the "smothered mate" happens. The ingredients are simple; a king who is totally entombed by his own pieces (usually in the corner of the board) is threatened by a knight who can leap over the defenders to challenge the king. Since the King has nowhere to run, a checkmate is the outcome.

A smothered mate usually requires sacrifice and a sequence of checks to compel the opponent to trap his own king, but to complete this example requires only one pass.

Bishop and Knight Fianchetto Checkmate

It is a common checkmate that exploits the 3 vulnerable dark squares around the castle of the Black King: the White pieces will occupy f6, g7 and h6. A bishop and knight may also work together to win a checkmate, but, either from their own pieces or from a few strategically positioned defenders, they can require a little more assistance to achieve so.

Exchanging the fianchetto bishop, who will be the dark square bishop of the Black, is often dangerous; it can produce several weak squares, especially if the opponent has a dark square bishop. If there is no extreme need to do so, do not substitute the bishop.

Anastasia's Checkmate

A excellent example of how the knight should be used in mating schemes is this checkmate. In the mate knight and rook of Anastasia, they work together to lock the king of an opponent on the h-file (it also operates on a-file, in the case of a long side castle) and then to checkmate. The knight controls the squares of g6 and g8, while the rook takes charge of the whole h-file and delivers a checkmate: while some beautiful combinations will lead to the final, in the above position, the mate is just one move away. White had just given the knight a check, forcing Black to play Kh7 in an effort to run.

Note: In the middle game, the mate of Anastasia reveals the value of a rook lift, which is a tactic where one player raises his rook from the back rank to an active position, usually close to the king of an opponent.

Two Bishop Checkmate

It was once said that, in an open position, a pair of bishops are twice as powerful as those bishops who are far from each other. Doubling the bishops, a very powerful weapon, is like doubling the rooks. Minor parts can also produce checkmates on their own. A pair of bishops can work well together; since each one can be dominant on squares of a single color, they can rule the entire board together.

Interesting Fact: Two bishops occupy 28 squares while operating at their best. The queen will do the same: 28 squares as well.

Queen And Bishop Pin Checkmate

The Queen fits well with a bishop in a pair. The g7 pawn did not catch the Bishop on h7 in the example above, as it was pinned to the King by White's Queen. That is a very simple and useful pattern of mating which should be regularly remembered and used.

Queen And Bishop Fianchetto Checkmate

That is another example of how it is possible to use "bishop less" fianchetto. White sees that and sets up the mating net cleverly. Squares h6, f6 and g7 are very weak and that fact is taken advantage of by White. With 1.Qg7#, White Checkmates. When coping with the poor, now you know what to do, "bishop less "fianchetto. For a queen, the bishop may play a supporting role similar to that of the knight in the preceding case. The bishop will support the queen from afar as the queen delivers the checkmate.

Note: Again, think twice before you swap your Fianchetto bishop!

Queen And Rook Checkmate

We do know that the queen and the bishop work together very well. We'll see in this instance that Queen and Rook function in pairs as well or perhaps better. Black has a side castle setup for his King, but the h-pawn is absent, making it a very insecure position for the King to be in. To produce a checkmate, our first example uses a queen and rook together. For any two major pieces, though, this same pattern can be done.

A lone king is easily verified by any two big pieces against the edge of the board. Although one piece stops the King from moving away from the edge, in order to have a checkmate, the other may move to the same rank or file as the King.

Note: It's a must to occupy it with big bits, double up and

strike when you have an open file available!

Queen and King Checkmate

It's a very easy but supportive playmate that almost always happens in the endgames. For a King and Queen working together, it is very straightforward to checkmate a lone King. The Queen is a strong attack piece, but in order to produce a checkmate, it normally requires some help. Many simple checkmates, assisted by a minor pi, use the queen to produce the checkmate. The queen is a strong attacking piece, but to produce a checkmate, it normally requires some support. Many simple checkmates use the Queen, supported by a small piece, to produce the checkmate.

Note: Most players are likely to leave until that position happens, but some players continue until the very end of the session. You must now, even though short of time, how easily and effortlessly to win these basic positions.

King and Pawn Checkmate

Even the smallest members of a chess army will take part in an enemy king's checkmate. Pawns can be very dangerous attackers in the right conditions.

There are only three potential ways for a king to get out of control:

- Move out of the way (though he cannot castle!)

- Bock the check with another piece or

- Capture the piece threatening the King.

Rook and Bishop Checkmate

This is a very general theme in checkmating, not just in the endgame, but also in the middle game. The basic principle is that the Rook is used to cut the Black's King on the side of the board and to deliver a mate using Bishop's long-range abilities.

Can I Checkmate with These Chess Pieces?

For beginners, even with superior power and the odds in your favour, it will always be hard to checkmate an opponent's king. Conversely, owing to the inadequate mating content rule, new players often attempt to play even after a game could already have been drawn. As soon as there is no way to end the game in a checkmate, the rule states that a game is drawn.

Here's a simple guide to what chess piece combinations you can (and can't) use when you're down to only two or three of your pieces against a single king to checkmate an opponent king. Or, maybe you're the one with the sole king, and this will let you know you can call a draw.

Checkmate Possibilities

Here is when you have the chance of checkmate and when it is a draw, with two or three pieces remaining. There are two or three pieces on the stronger hand, while the weaker side only has a king:

• King and queen versus king: The stronger side should be able to checkmate easily.

• King and rook versus king: The stronger side can checkmate, and while it may take more moves than with a queen, the technique is still quite simple.

• King and bishop versus king: The stronger side cannot checkmate.

• King and knight versus king: The stronger side cannot checkmate.

• King and pawn versus king: The stronger side may be able to checkmate, depending on the position. The goal here is to promote the pawn into a queen, after which checkmating is fairly simple.

• King and two bishops versus king: The stronger side can checkmate, though the method for doing so is somewhat more complex than with a rook or queen.

• King and two knights versus king: The stronger side cannot force a checkmate, although it is possible with the cooperation of the enemy king.

• King, bishop, and knight versus king: The stronger side can checkmate. However, the technique is rather difficult, and even many strong players have failed to properly convert this endgame, especially in time pressure.

• King versus king: This endgame is always a draw, despite the heroic efforts of beginners and scholastic players who have danced their kings around the board for dozens of moves.

Of course, several checkmates take place in these cases with even more information on the desk.

There are two explanations for the better use of this knowledge. Second, in hopeless circumstances, it will deter you from continuing: if you are down to a king and bishop versus an opposing king, for example, it is time to start a new game. Secondly, when you have a material edge, it helps to consider what endgames you can go to that would win. If you have a bishop and two pawns vs. a bishop, if you can later promote a pawn and win that way, it's good to exchange those bishops. Allowing the other player to exchange their bishop for your two pawns, however, would result in a draw.

Five ways to get to behave like a Grandmaster

There is one thing a devoted player can accomplish, regardless of the level of play or power, and that is the capacity to "think like a grandmaster." We frequently struggle with issues such as poor calculating ability, lack of planning for opening, no understanding of the endgame or even managing the time pressure that keeps us from achieving a higher level.

This does not mean, though, that we are far from a Grandmaster's reasoning process; it just suggests that GMs tend to overcome these problems more quickly and more frequently than we do.

Like a Grandmaster, how to think?

What is it that the majority of us see differently? We're going to give you 5 tips in this article that could help you improve your chess thinking to a higher level and thereby making better choices on the chessboard.

Needless to mention, the attitude has nothing to do with the shape of the game. In all aspects of chess, each GM has a fair amount, some better than others, but everything adds up in the end and that is what makes a GM a difficult player to beat. Because of his overall level, which can only be reached by preparation and playing, he can solve tactical, positional and endgame problems.

What Would A Grandmaster Do?

1. Look deeper, do not get distracted by "cheap" strategies

Quite frequently, the lesser players are quickly tempted by fast opening attacks or even middle-game strategies that are not so hard to find. The notion of "but maybe my opponent won't see it" needs to be absolutely erased from your mind; with it, you're going to lose more than you win.

A GM is not interrupted by this. In fact, the last thing he's going to try during the game is strategy, and that can happen because of two things. Either there is no way to deter them, or they seem promising. Look for deeper motions, sound moves that strengthen your position and generate subtle issues for your opponent; one of the main aspects is that.

2. Evaluation of The Position

Believe it or not, in order to gain excellence in chess, this is probably one of the most important things to learn. Any of the options derive from a prior appraisal of the role you are playing. A incorrect assessment will apply to a poor idea and vice-versa. You need to figure out what really is wrong with it and aim to change it if the situation is worse.

You need to work out what kind of profit it is and how to

make the best of it if you have an edge. If the situation is equal, wait for your chance and let your competitor take the gamble if you keep making changes or stagnant moves that do not deteriorate the position. When playing with weaker opposition, many GMs apply this tactic. They reinforce their positions and "wait" for the opponent to launch the first attack (wrongly).

3. Overestimation of One's Position

As you enjoy attack, learn to understand defence. The art of attack is so much rightly celebrated, and we overestimate our chances when we face positions that look nasty or are impossible to defend for our adversary, assuming that victory is inevitable and yet it is not. Open your eyes to the position and have no bias. You could be stronger, but not enough to succeed, and that's OK.

Trying to win it at all costs is one thing you can't do. We over-press, deteriorate our own position often, and then the result is determined against us. We should always be mindful of the chances of our rivals, especially when they are worse!

4. Emotions

Emotions, indeed. For any chess player, emotions are a great opponent, but when you lack experience, it is much worse. Typically, they emerge when you score, and that's the hardest part. You're curious about what's going on the board, you're already starting to envision a post-mortem examination, and suddenly the win goes farther and farther down, and it slips away more often than not.

A GM at all times holds his blood cool. This is good because it just retains rational reasoning and focus on what's going on on the board. Don't get swept away on the chess board with passions. As if you were completing a school exam, try to keep it simple; simply answer the questions (play the right moves).

5. Self-confidence

Casablanca once said that you ought to play it if you see a good pass. When you believe in it, just play it; it doesn't matter if, in the end, you lose. The point is that based on your judgment can you base your decisions.

If you feel like the job needs risks to be taken, go ahead. Do not be threatened by your opponent's strength or by what your coach or training partner may tell you after the game. Just trust the logic behind your moves and go on, if you loose your learning as well.

Before they become normal to you, you may want to read these ideas once or twice. It would surely make a difference in the way you see chess and play it. Thank you for reading and feel free to comment with your opinions and tips, as always!

CHAPTER SEVEN

Getting Better at Chess Openings

Research means practice, and it is just impossible: you have to do some work if you want to improve your play seriously. This chapter will help you get the best out of your studytime rather than work smarter.

Some people expend a great deal of time researching openings for chess. Right now, someone is researching a chess opening in order to change one of its variants somewhat. You don't have to be that guy.

Take the time to study the openings, but focus on mastering the thoughts, not memorizing the gestures. You will still discover a pass you have never seen before if you play enough chess. You would have a solid foundation for a decision of how best to confront the unexpected once you grasp the concepts behind an opportunity.

Get a coach

In 1972, I couldn't find chess coaching when I first began to play tournoy chess. They are regular nowadays. Of course they differ greatly, but you will almost definitely find what they are responsible for their time and how successful they are.

You should still get an Online video phone service to find a coach on the Internet if you can't find a coach in your vicinity. Face-to-face lessons are best, but not terrible at all telephone lessons.

Good coaching can test not only your openings awareness, but also your drive and commitment in improving. You customize your lessons to fulfill your desires. You will need a new coach if you don't do this.

Nothing is better than a good player who can explain the thoughts behind an opening patiently. He or she should respond to your questions and explain that one shift is preferable, even though the shifts are fairly the same.

Searches for a mate

A less formal approach to grasp the opening of chess is to find a mate who understands more. Any chess players have their expertise, but most of them are more than ready to step up and demonstrate their brains.

Ask your buddy about an opportunity he or she knows about and be willing to ask questions. Often people believe that even though they aren't those things are evident. Talk up if you don't know what your buddy means to you!

At least when you believe you know what your friend is all about, play those games. You certainly will make mistakes, and you will probably fail, but with the opening you have chosen to practice, you will gain experience.

There is no substitution for experience, and the way to achieve experience is no easier than by playing chess with a friend. Maybe for a while you have to grant bragging rights, but you're going to get stronger!

Annotations Reading

If players take records of movements in a given chess game, they are known as annotations. Playing well-noted games is an excellent way to practice openings.

Any players' annotations are only different from the current game. These aren't going to be very effective at first. Check for games in annotations that have more prose than moves.

In books or online you can find well-noted games. Of course, you can't ask a book any questions, but you often find sites online for clarification or consultation.

Removal and drag

Before the Internet made discovering other chess players too easy, I got serious about chess. We actually played chess on postcards with people elsewhere. One player will send another pass and then wait to get an answer from the return mail.

It was sluggish, but it was an excellent way to learn new opportunities because I'd put the game in a board and start from scratch anytime I had a postcard from my opponent. I could see the opening moving on again and over before I could recognize them. I could see the opening shift again and over.

Today, you will play this kind of chess on the Internet with people worldwide. Web servers allow you to drag and drop a piece on a square, and the move is automatically redirected to your opponent. You may not have to answer your adversary for a couple of days, but the transmitting time is not there. None against the United States. Postal, but certainly the mail beats!

Many websites encourage you to back up a game and move it one stage at a time. When you learn a new opening, I suggest that you do this. The repetition has a permanent influence. Ideally you place the game on the real board and move the pieces physically around, but it may just be too much in this era of ease.

True Time Play

You have an award of time (called "time control") when you play chess in real time (in contrast to correspondence chess) when you have to complete or forfeit a game.

It is a more pressure-packed way to play Chess, but it is thought by others to stimulate addiction. When my wife came home from work, I had to abandon one site cold turkey, and I knew that I had wasted the whole day online chess play!

I just recommend in balance in real-time chess, but this is a way that you can gain a lot of practice at any opening that you want to play. Various players reaction to your movements differently and you will easily build an outline of the ideas involved.

It's blinking

You should try what is called blitz chess if you are familiar with chess clocks and have a friend who wants to play short games. You have a very fast time management in blitz chess. Both players can only play the whole game for 5 minutes. You would actually flash your gestures up!

Blitz games can be very chaotic, and some people just don't like them. However, if you do, the quickest way to acquire a lot of experience is a lightning game. You don't have time to slow down and ponder, and you must learn to understand the patterns, and quickly pick your movements.

Blitz's not for everybody, and surely it's not for heart faintness. It's a decent way, though, to measure how well you know an opportunity.

Basing the Data

Chess games were compiled in databases which basically contain millions of games. They can be grouped into various openings and the corresponding tools can be used as fast or slowly as you choose to play with these games.

Some are free of charge, but others need support to be available. The free sites are more than enough for all players except the most committed.

Revving an Engine

Chess playing machine applications are known as engines. They came a long way, sweetheart. In 1985, I played a two-game match against a Cray-super-computer software developed by Bell Labs. I won and lost one game. I may purchase a better application on my PC ten years back.

You may use these engines for the study of the chess openings if you don't have a coach or a good chess player to call. You can play video games in the background or do real work and it gives you a reasonably clear idea if the opening is better for White or Black. They score the position numerically — a positive number means that White is fine, and a negative number means that Black has the advantage.

Studying Your Games

The best way to learn Chess, particularly the games you have lost, are to study your own games. Using chess to document every serious game you are playing (if you play online, the moves are usually recorded automatically).

Try to be as purposeful as you can. Identify when and why you have gone wrong. If you were mistaken, find out for the next time an upgrade. If you have discovered it on your own you certainly would have no problem recalling the change.

It never hurts letting a better player or engine evaluate your playing, but it makes the deepest impact on yourself. It might be more fun to reveal your gains, but if you research your losses, it will pay larger dividends.

Book buy

More books on chess openings have been written than on any other part of the game and the trend will possibly continue.

Although some are highly technical, all of them are starting point-oriented. By searching around, you can find the best match for you. You will decide that the kind of book you need a certain chess editor is right for you; or it may only be a ticket for a certain author.

You're going through a trial and mistake and it may not be right for you what's right for the next guy, but there's a massive ocean of chess literature. Don't be scared to dip into it your toe.

Mistakes to Avoid in the Opening

In chess, mistakes are possible. It's all made by everybody. It's yet to play the perfect game. A wagon once said that the chess champion is the player who makes the last miss.

However, you want to minimize the number of failures, and you want to benefit from them if you do. When you figure out that you have lost a game in the future, you play better. In this chapter you can find some general thumb rules that will help you discourage cardinal sins from coming into an opening, or help you at least remember them if you commit them.

Wasting Time

All movements are important. It's the same as your enemy jumping twice in a row. This isn't a success formula.

Each move at the opening should be a progressive move, one that increases your pieces' versatility. The strength of the parts is related to their mobility. The most mobile it is, the better it is.

You will ultimately be outgunned if you don't make developments, however your competitor does. A few time-consuming techniques are present to prevent:

• Move the same piece numerous times: while you need several openings to move the piece more than once, you can do this for a good cause and if the majority of your pieces have still not been created.

For instance, in the Sicilian, it is totally reasonable to move the knight after playing Nxd4 for a second time, after moving 1.e4 c5 2.Nf3 Nc6 3.d4 cxd4. The move recovers the pawn on d4 and maintains the balance of material and centralizes the rider.

However, playing 3.Ng5 will be an error instead of 3.d4. This second knight move wastes a chance to improve the remaining bits. The bishop of white opens the line for 3.d4, but 3.ng5 still doesn't encourage the bishop to pass. 3.

Tell Checking: Checking is not an end by itself, but certain players are checking the king of their opponent any time he has. It's a waste of time sometimes. For instance, playing 2.Bb5+ will be waste of time in the Scandinavian after the moves 1.e4 d5 due to the response from Black's 2. c6.

The bishop must move again after 2....c6 or the pawn avoids being hit. It is Black's turn again after the Bishop's going, and move 2....c6 is cast free of charge.

Leading with the Lady

Most players like to transfer their Queen early on. Typically, this is an accident. The queen is the best piece of chess, so that means that you can carefully grow your queen.

If the queen is deployed too early it will allow less useful bits of abuse. This harassment forces you to lose time by changing the queen numerous times, and worse, it could be that the queen is missing for less than that.

It's a mistake to use the queen to conveniently parrot attacks. For instance, white queen threatens to catch Black's pawn on e5 after 1.e4 e5 moves 2.Qh5. But Black will respond by 2....Nc6, 2....d6, or 2..Qe7 to the danger. After the natural progression of Black's Ng8-f6, the White Queen would be targeted and forced to withdraw.

The moves of 1.e4 e5 2.d4 exd4 3.Qxd4 Nc6 indicate the pitfalls of developing the queen too early at the opening known as Center Game (see Figure 23-3). Black's knight is well positioned on c6 because it takes time there when the queen of White has to withdraw.

White will start the game with a time advantage because of the first pass. It doesn't make sense to waste this profit when the queen grows too early.

Losing Material

You start the game with an equal battle power. You have made an error if you loose all of your power without any payout. You have to maintain the balance of materials or you risk losing the game.

If you sell information to your rival, be careful — you might capture it. However, if you do not see any adverse impact after careful consideration on taking the material given, take it by all means! The best gain to convert into victory is a preponderance of force.

Often you're not going to lose the stuff immediately. The other shoe most much drops after your king or queen is attacked.

In the middle White seems to be a lovely edge. White has three pieces created, while Black has only one. However, by fighting the queen with ...c5 Black is able to gain stuff. After the queen leaves d4, so c4 is stuck and the bishop is won by b3. The bishop's value is over three pawns, so Black gets stronger and White comes close to the loss a couple of moves down.

Abandoning the Center

You finally occupy more room than your competitor if you control the centre. More room helps you to maneuver more items. Greater handling brings more strength to them. You can find that you can compel your adversary to make sacrifices that maximize your benefit so that you will win.

Anything because you managed the centre.

However, you make a mistake when you quit the centre. Please remember position 1.e4 h5 2.d4 h4 3.Nf3 a5 4.Nc3 a4 after movement.

Knights of White have been created to better monitor the centre. The Bishops of White are able to evolve. It is hard to challenge white's dominance of the centre. Black's pawns make nothing concrete and Black's pieces find sufficient posts challenging. At least some stake in the middle, otherwise you are in danger of being overwhelmed, is desperately necessary.

Creating Weaknesses

When all pawns advance to adjacent files, squares can be substantially weakened. If a pawn cannot guard a square, this is an open invitation for the pieces of your enemy.

White also opens with 1.e4, a really nice move. But the advance of the e-pawn has a small downside since the pawn can't be used anymore to manipulate the squares d3, d4, f3 and f4. The move has much more strength than disadvantages, but its disadvantages can be accentuated if White still supports the g-pawn.

You will have to cover them with pieces if you have poor squares in your place. Parts do not work in the execution of certain guard duties, so it is best not to do so. Of course, it might be worth it if you can win a queen at the risk of having a weakness! However, it is best not to build vulnerabilities in the first place, all things being fair.

Tombshell Pawn

Pawn grabbing is the term used in order to explain how mistaken you took so long to get a pawn. Chess players talk of a "poisoned" pawn, which means it doesn't worth the material reward the time they waste capturing it.

A line in the London system helps Black to pick up a pawn before movements 1.d4 Nf6 2.Bf4 c5 3.e3 Qb6 4.Nc3.

When Black captures 4....Qxb2 pawn, so 5.Nb5, which threatens 6.Nc7+, will be played by White. White can drive the draw by hitting the Black Queen twice, who has no way of escape, or White can follow the victory. You don't want to make that kind of decision too early in the game to your rival.

Exposing the King

It is an error to subject the king to needless harm. Most opportunities require early castling so it is too necessary to keep the king secure. In certain ways, the protection of the King is weakened to guarantee a certain kind of gain, but those ways are dangerous.

The motions 1.e4 e5 2.f4 exf4 3.d4 Qh4+ are one modification in the King's Gambit. White needs to move her when 4.g3 fails to 4....fxg3. White lacks the opportunity to create just a solid cobblestone centre, but experience has proven this tactic to be imprudent.

Blocking Lines

Often a diagonal must be blocked to position the argument in the middle. This occurs in the French Defense as well as the weakening Queen Gambit. However, it is an error to block a line for no reasonable cause.

Typically White needs to support the d4 pawn by playing e2-e3 as he moves 1.d4 d5. But to do so right away will block the bishop's line from the dark square.

Few opens use this technique, such as the Colle Method, but normally before playing the e2-e3 it is best to improve the bishop to f4 or g5.

Falling for Traps

Opening pits are an indubitable, but considered to be poor series of steps. You must know if pitfalls are involved when choosing an opening for yourself.

For eg, if you want to play Petroff's Defense, you must be mindful that capturing 3....Nxe4 is a mistake after moving

1.e4 e5 2.Nf3 Nf6 3.Nxe5. White is playing 4.Qe2 in the lead.

Black can play 4....Qe7, but after 5.Qxe4, White can't copy the step.

The knight in White is now safeguarded. Remember that the Knight cannot retire with 4...Nf6 on e4 because 5.Nc6+ wins the Queen of Black!

Memorizing Moves

It is a failure to memorize gestures without learning the theories behind them. You need a certain amount of time to say one another's opening and escape traps, but afterwards it is much more important to know the ideas behind a given opening than to know several variants that pass profoundly.

You want to know the general strategy behind a certain opening and the general concepts of the opening game. If you know any traps that you want to open, you can avoid any tragedy during the release.

Becoming a Better Chess Player

1. Learn how to play: If you do not know the rules or how to move a piece correctly, you can't get better.

2. Join a nearby chess club: Chess can be cooperative and open. Don't make yourself feel good by playing alongside persons that are obviously inferior than you. A smart way is to start preparing how to counter your adversary if you have to make yourself feel better after a defeat.

3. Learn the values of the pieces: One point is worth a pawn. Each of the Knights and Bishops is worth three points. It's worth five points to A Rook. Nine points is worth a Queen. This is just a guideline, not a winning technique, so you can ignore the piece values if you have a forced victory on your turn.

Don't unnecessarily give up material: While a well-planned sacrifice will often place you well ahead in the game, missing pieces will do just the reverse due to bad planning. Defend the pieces well and consciously prepare compromises.

Trading a Bishop (worth 3) and a Knight (worth 3) for a rook (worth 5) and a pawn (worth 1) is not desirable since the Knight and Bishop are more powerful than a rook and before the very end of the game the pawn will not come into action.

• The values are relative. A bishop or knight is better in some positions than a rook.

Despite its obvious importance, an interchange (a knight or bishop for a rook) is NOT worth 2 points. Usually, it is worth 1-1 1/2 points. Therefore, 1-2 (sometimes 3) pawns are sufficient consolation for an exchange being down.

4. Bishops and knights still grow. Pawns are overused and overextended, and the parts that develop sometimes do not develop. Then, the enemy will normally bring a bishop through the structure of your pawn.

Moving too many pawns weakens the side of the castle king and opens you up to attack. Typically, moving too many pawns will disrupt the pawn structure of your endgame.

5. Find your playing style. There are several different ways in which individuals play chess. Some are easy to initiate attacks, play gambits, or make concessions, and enjoy offensive playstyles. Others favor silent strategic play, usually taking several turns setting up a powerful position before any attack is launched. Check out different playstyles and learn what you prefer.

6. Get the first tournament. Go there feeling like in this series of games you are going to pound ass. Forget about the ranking. Forget the ratings. It is a self-fulfilling prophecy to

just get out there and play the best you can.

7. Get a rival. Pick somebody and "compete" against them who is stronger than you. Oh, play them. Go to the competitions they're attending. Get used to their playing style slowly and use it against them and other individuals. Don't think about this "rival" as someone who can do better than that. When you lose, don't beat yourself up. Again, play them. And that again. And that again. Until you have mastered their style and how to fight it, do this.

8. Study your GM favourite (grandmaster). Studying, playing, studying, playing. Study how to use, and how to counter, their tactics.

9. Study the rules of the simple endgame. "Endgame strategy, "Exchange pieces not pawns if ahead in material. If behind in material, exchange pawns and you will force a draw.

You must be at least a rook up to force mate without pawns, with the only difference being that two knights and a king will not force mate against a lone king.

The king is a strong piece which can be used to block and strike pawns.

Bishops with opposing colors draw much of the time and without destroying them, neither side will advance pawns. If the bishop is the opposite color as the queening square, a rook pawn and bishop draw only against a black king.

Bishops in all but locked pawn positions are worth more than knights.

If the game continues, pawns, rooks, and bishops become more important and play to hold them.

Many games finish in a tie with all the pawns on one side of the board. 90% of master games end in a draw because all the pawns are on one side of the board and the master can swap pawns for the fewer pawns and then sacrifice with the last of the pawns a knight or bishop. You will not force mate if you are left with only a Bishop or Knight.

Rook and Knight, or Rook and Bishop, will only draw against a Rook several times.

In Queen endings, traditionally, the player who first pushes the Queen to the center dominates play.

10. Powerful Pawn Structures are:

A "Outside Pawn" draws the king of the enemy to the other side, allowing you to swallow the rest of his pawns on the other side of the board or advance your pawns.

A "Passed Pawn" is not obstructed by another pawn and can be moved. "Passed Pawns must be pushed" said

Nimzovitch.

A "Protected Passed Pawn" is a passed pawn that another pawn defends. A Protected Passed Pawn requires the opponent to defend against an advance on a continuing basis.

11. Weak Mechanisms of the Pawn are:

• Doubled pawns cannot defend each other and are subject to attack.

• Isolated pawns are weak and must be defended by a piece.

• Backward pawns on open files are extremely weak and subject to attack by rooks.

• A King with the opposition can draw against a King with a Pawn.

• A Rook on the seventh rank is worth sacrificing a pawn.

• Zugzwang is where if your opponent moves his position becomes weaker (he would rather give up his turn), and is common in Chess.

• Rook and Pawn endings are the most complicated so avoid them.

12. Play blindfold chess: It will train you not to forget and relearn which pieces strike which squares before you look

and see. Because the brain will already be forced to memorize too much knowledge about the condition of the game, it will not be much harder for it to learn to arrange the information it knows about the board into a different set of information pieces than to learn to organize it.

13. Notice patterns of what moves appear to help you win the game: don't necessarily follow step 3 without deviations, but rather judge the arrangement of pieces and determine if it's actually worth making a deal. If you're ahead as seen in the following statement, it's nice to trade pieces more quickly. If you had a forced win if you promoted a pawn to a rook that would entail losing the rook, you would have a forced win

TOURNAMENT RULES

When you are not aware of the rules of the tournament, what is the point of learning about the different implementation techniques in chess? If you are left in the dark on tournament rules, the odds of you losing the game are very likely. This is precisely why I have mentioned for your comprehension the essential tournament rules in this chapter. Enable us to look at them one at a time, and by the time we're done, I'm sure you're all set for a real chess game.

When you're in trouble, Ask

There is nothing wrong with the tournament director clarifying the questions. If you are not very confident about a certain tournament rule, make sure that the tournament director clarifies your questions before the game starts. This way, if you wish to ask these questions in the middle of a game, you can save time that can be lost.

Similarly, if you and your rival disagree about something or any rule while the game is in progress, the smart thing to do will be to pause the clock and ask the tournament director to explain it for the two of you. This way, by indulging in meaningless talks, you and your rival would not lose time.

You Have to Move the Piece You Touch

This is otherwise referred to as the law of touch action. Going by this rule, the piece you directly strike will have to be moved. Especially if you are a novice, this is a rule to watch out for. With your bits and moves in a tournament, you should not keep fiddling. Therefore, there is no undoing a sloppy pass, so be careful of which piece you hit.

However, should you strike a piece by accident while reaching out for another piece, this rule should not be instigated by your opponent. You are not allowed to move the piece in order to unintentionally strike it. "If you think you need to adjust a certain piece because it is not correctly placed on the board, you can do so before touching the piece by saying, "I adjust. This way, this maxim will not be invoked by your adversary to make you move the piece you just touched. You are making it clear when you say this that it is not your intention to contact the piece to move it.

Recording The Moves

In most tournaments, this is another general law and a few tournaments do not have this policy. In most tournaments, players are required to write down their moves and the purpose of this is that it can serve as a written record of what happens during the game. In the event of the need to settle any conflict that might occur later, these should be referred back to. Therefore, it should be performed with full focus to log the movements. Learn how they can be written down. If you do a poor job of recording your moves, when a disagreement happens during the course of the game, you may not be able to use it to your benefit.

Do Not Interfere With A Game

Usually, players who don't have a scheduled match are able to wander around and watch how the other games are going. Recall that your position is confined to that of an observer. In other words, at some moment in time, no matter what the justification is, you should not intervene in the game. For example, it is not your duty to report it if you see a certain player making an illegal move. Let the players play their way around the game and don't meddle with it.

Similarly, you are not allowed to provide another player with feedback or hints or tricks. When the game is going on, you cannot support any player. For doing so, you can also get disqualified from the competition. Therefore, for no cause, they never mess with an existing game.

Remember To Turn Off Your Cell Phone

Nothing like a mobile phone that rings loudly in the middle of a game can be too annoyingly irritating. The bulk of these games, as we all know, are clocked. This loud phones will, thus, quickly interrupt players and waste their valuable time. This is why, as the games are going on, tournament rules have been changed to restrict the use of cellular phones. According to this regulation, you will be punished with a penalty if your phone rings in the middle of a game. And if it was the worst scenario you thought it was, then you're mistaken. Mobile phone use can also lead to the game being forfeited. Consequently, before the tournament starts, make sure that you turn off your computer.

Learn to Use the Chess Clock

This tournaments are, as I said before, bound by time limits and you are expected to finish these games within a defined time period. Using these clocks while you are a beginner can be a challenge, but with ample practice and time, with the aid of the clock, you can find it easier to play the game. You will find that in keeping with the timer, the brain will start running. As soon as you have made your pass, hit the clock as this is to guarantee that you do not spend your precious time on nothing. Note that with the same hand you use to move the pieces, you must strike the clock. Therefore, make sure you make it a point to improve the coordination of this hand a few days before the tournament.

Record Your Result

Make sure you report the outcome of the game as soon as the game is over. It is all players' responsibility to report the game's outcomes. If you are not sure how to report your scores properly, then call your tournament director to help you do it.

In the majority of tournaments, these are some of the essential rules. Some tournaments often have extra rules to guarantee that you are familiar with the various rules of the tournament before it starts. If you struggle to do this, the numerous rules declared before your game starts will leave you at a great disadvantage and in a puzzled state.

CONCLUSION

Now that we have come to the end of the Chess Opening for Beginners guide and maybe you have already tried other guides in the past, but you just could not find a suitable opening for you. This time will be different.

Again, let me thank you for reading my guide. There are a number of great books on the topic, so I really appreciate you choosing my guide. If you enjoyed the book, I'd like to ask for a small favor in return. If possible, I'd love for you to take a couple of minutes to leave a review for this book on Amazon. Your feedback will help me to make improvements to this guide, as well as writing books on other topics that might be of interest to you. Hopefully this will allow me to create even better guides in future!